Penguin Education
Penguin Library of Physical Sciences

Organophosphorus Chemistry

B. J. Walker

Errata

Page 21
Table 5, heading to third column *should read* *Bond length*/10^{-10}m
line 3, *for* 0·19 × 10^{-11}m, *read* 0·19 × 10^{-10}m
line 4, *for* 0·30 × 10^{-11}m, *read* 0·30 × 10^{-10}m

Page 34
line 6, *for* 2·5 × 10^{-12}m, *read* 2·5 × 10^{-11}m
line 12, *for* 1·39 × 10^{-11}m, *read* 1·39 × 10^{-10}m

Page 35
line 4, *for* 1·65 × 10^{-11}m, *read* 1·65 × 10^{-10}m
line 5, *for* 1·60 × 10^{-11}m, *read* 1·60 × 10^{-10}m
and *for* 1·80 × 10^{-11}m, *read* 1·80 × 10^{-10}m

Organophosphorus Chemistry

B. J. Walker

Penguin Books

Penguin Books Ltd, Harmondsworth, Middlesex, England
Penguin Books Inc, 7110 Ambassador Road, Baltimore, Md 21207, USA
Penguin Books Australia Ltd, Ringwood, Victoria, Australia

First published 1972

Made and printed in Great Britain by
William Clowes & Sons Limited
London, Colchester and Beccles

To Margaret

Contents

Editorial Foreword

For some years, it has been obvious to most teachers of organic chemistry that the day is past when one author, or a small group of authors, could write a comprehensive textbook of organic chemistry suitable for the courses in this branch of the subject as taught in universities. While a number of excellent textbooks are available, they must now almost of necessity be either incomplete in their coverage of the subject or massively unwieldy and correspondingly expensive.

It seemed to the publishers and to the editor that a useful method of coping with this difficulty was to present the material as a series of smaller texts on various aspects of the subject; by producing them in the format used in this series their price has been kept down to a minimum. In the books on organic chemistry, those parts of the subject which are generally regarded as essential and common to all universities' degree courses will be covered, as also will many topics which appear as optional subjects in many degree courses. It is hoped that undergraduates will wish to purchase such of the volumes as are relevant to their own particular courses.

The books in the series have been planned essentially as undergraduate texts although it is also hoped that they will be of use to more advanced students who require concise introductions to particular aspects of the subject. Since each book has been written by experts in its subject, it is hoped that all the volumes will come to be regarded as authoritative texts. It is also hoped that undergraduates and other readers will appreciate the inclusion of some worked examples and problems in many of the books.

G.H.W.

Author's Acknowledgements

It is a pleasure to express my thanks to Professor H. B. Henbest and Professor S. Trippett, both for reading my manuscript and for many helpful criticisms. I should also like to thank Professor G. H. Williams and the editorial staff and others at Penguin Education for their skill in preparing this book for publication. Finally, I am especially grateful to my wife, Dr Margaret Walker, who, in addition to reading the manuscript, is the author of Chapter 4.

Introduction

Historical

The first preparation of relatively pure elemental phosphorus was carried out by Hennig Brand in 1669. Brand had been trying to repeat some work which purported to convert silver into gold and, during his naturally unsuccessful attempts, had cause to distil large volumes of human urine. One of the final products of this distillation was a liquid which 'emitted a strange light'. Brand was able to obtain solid samples from this but found that, unprotected from air, these lost the ability to phosphoresce fairly quickly (through oxidation to phosphorus oxides, although of course Brand did not realize this). Brand was a rather gullible character and in a moment of weakness shared his secret with a well-known seventeenth-century alchemist, Johann Kunckel. Standards of honour were obviously not very high amongst chemists of the time and Kunckel allowed it to be known that *he* had discovered the wonderful 'cold fire'.

Another distinguished contemporary chemist, Gottfried Leibniz, became involved with Brand and, since Leibniz was influential with the Duke of Hanover, Johann Friedrich, he and Brand were able to gain access to a large urine supply (about a hundred tonnes) from the local garrison. Unfortunately the Duke died and the large-scale preparation was never attempted. Brand had also been promised money by Leibniz and when this was not forthcoming they never spoke to each other again.

In England, Robert Boyle came to hear of the discovery and obtained some vague details from which he was able to repeat the preparation and publish it in 1680. Boyle died in 1691 but his assistant, a German immigrant called Ambrose Hanckwitz, was a shrewd man and began to produce phosphorus on a relatively large scale. He managed to capture the market for almost fifty years, and even in 1731 phosphorus cost two pounds sterling an ounce because nobody else was in a position to supply it.

Brand and his contemporaries believed that they had prepared the element 'fire', and early references to elemental phosphorus refer to this and not to the modern chemical sense. This concept slowly disappeared and, in the late eighteenth century, phosphorus played an important part in Lavoisier's theory of combustion since, together with sulphur, it provided the experimental evidence of increase of weight on burning. These experiments finally led to the recognition of phosphorus as a chemical element.

In the early nineteenth century the first really commercial preparations of phosphorus were begun by a variety of small manufacturers; in 1851 Arthur Albright, the founder of Albright & Wilson, began to produce phosphorus on a more ambitious scale. This process involved the reduction of phosphate ores with carbon in sealed containers, followed by distillation of the phosphorus formed. All the phosphorus prepared at this time was the white allotrope and because of its pyrophoric properties the great majority was used in match manufacture. Unfortunately, prolonged contact with white phosphorus leads to an extremely dangerous and unpleasant disease known as *phosphorus necrosis*, or 'phossy jaw', which involves the formation of discharging sores, particularly around the jaw, followed by destruction of the jaw-bone and death. When red phosphorus (which is relatively harmless) was discovered in 1847 by the Swiss chemist Anton Von Schrötter, it soon replaced the white allotrope in matches, although in some countries (for example, the USA) white phosphorus was used well into the twentieth century.

The original commercial process for preparing white phosphorus was very expensive and by about 1900 had been replaced by a continuous process involving electrically heated furnaces. This process was in principle the same as that used to prepare white phosphorus today, although in present-day phosphorus plants extensive safety precautions are in operation and 'phossy jaw' is a virtually unknown disease.

By the second half of the nineteenth century the first organophosphorus chemistry was carried out and in the 1880s and 1890s a considerable amount of work was done, particularly in Germany by A. Michaelis. However, it was not until the 1920s that organophosphorus compounds had any commercial value. The real expansion in their use came in the 1940s as additives in the rapidly growing plastics industry.

The commercial importance of phosphorus and its compounds

In terms of weight, the great majority of phosphorus compounds used commercially are inorganic. The fertilizer industry accounts for about 70 per cent of the total phosphorus used, largely in the form of inorganic phosphates (e.g. 'super' phosphate, $CaSO_4$ and $CaHPO_4$). Detergents and animal feedstuffs account for 15 and 8 per cent respectively, while pharmaceuticals, insecticides and plasticizers, which use largely organic compounds, involve only about 2 per cent of the total phosphorus. However proportions of the market do change, for example over the next few years there is certain to be a large drop in the use of tripolyphosphates in detergent manufacture, since these constitute a pollution hazard. Through drainage etc., virtually all detergent phosphate ultimately finds its way into rivers and lakes where a high phosphate content promotes marine plant growth, often to an alarming degree. This leads to the blocking of rivers and canals and a disturbance of the natural balance of animal and plant life. Since detergents consist of approximately equal quantities of polyphosphate and a *surfactant* (usually sodium alkylbenzene sulphonate), a vast amount of phosphorus is being added to the natural cycle in this way.

Although the actual volume of organophosphorus compounds used commercially is small, their number and importance is disproportionately large. The oldest large use is in plastics where they act as antioxidants and stabilizers. Organic phosphates ($(RO)_3P{=}O$) are also effective plasticizers (increasing flexibility and working qualities) and act as flameproofing agents. Phosphates are also used as petroleum additives to control pre-ignition and in hydraulic fluids, particularly in aircraft because of their fire resistance. *O*,*O*-dialkyl phosphorodithioates

$$(RO)_2P(=S)SH$$

are used in lubricating oils, since they are extremely effective in reducing wear and will still lubricate at high pressures, where hydrocarbons are ineffective. Organophosphorus insecticides (see Chapter 5) are widely used in agriculture. Their use is likely to increase as they are fairly quickly decomposed in the soil and so less likely to cause damage to wild life than the halogenated hydrocarbons.

Virtually all the organophosphorus compounds used in these ways are prepared from elemental phosphorus through the intermediary of phosphorus trichloride PCl_3 and phosphoryl chloride $POCl_3$.

An organic chemistry of phosphorus

The chemistry of organophosphorus compounds is made up of two types of contribution. First, there are the reactions of the phosphorus atom itself, and then there are the ways in which the presence of the phosphorus atom modifies the reactions of the organic substituents.

Phosphorus, like carbon, occupies a fairly central position in the periodic table and so can easily form bonds with both electronegative and electropositive elements. These bonds are generally strong (e.g. P—O, P—F, P—C; see Chapter 1), and so a wide range of phosphorus compounds exists. In the trivalent state, phosphorus possesses a lone pair of electrons and is readily polarizable, which enables it to act as a nucleophile towards a wide range of atoms (see Chapter 2). However, phosphorus is relatively electropositive and so can also act as an electrophile. This behaviour is assisted by the ready conversion of phosphorus(III) to phosphorus(V), which further increases the range of compounds obtainable.

The stability of the pentavalent state is probably due to the involvement of 3d orbitals in phosphorus bonding (see Chapter 1), an involvement which can be used to explain many of the more unusual aspects of phosphorus chemistry. For example, the stability and high bond energies associated with compounds like phosphonium ylids ($\geq P^+{-}C^-\langle$), imidophosphoranes ($\geq P^+{-}N^-{-}$) and phosphine oxides ($\geq P^+{-}O^-$) are most easily explained in terms of 3d-orbital participation in π-bonding. The involvement of 3d orbitals also increases the range of phosphorus radical chemistry (see Chapter 6) by stabilizing radicals with more than eight electrons about the phosphorus atom ($R_4P\cdot$).

Table 1

	Group IA	IIA	IIIB	IVB → IIB	IIIA	IVA	VA	VIA	VIIA	Inert gases
	H $1s^1$									He $1s^2$
$1s^2$	Li $2s^1$	Be $2s^2$			B $2s^22p^1$	C $2s^22p^2$	N $2s^22p^3$	O $2s^22p^4$	F $2s^22p^5$	Ne $2s^22p^6$
$1s^22s^22p^6$	Na $3s^1$	Mg $3s^2$			Al $3s^23p^1$	Si $3s^23p^2$	P $3s^23p^3$	S $3s^23p^4$	Cl $3s^23p^5$	A $3s^23p^6$
$1s^22s^22p^63s^23p^6$	K $4s^1$	Ca $4s^2$	Sc $3d^14s^2$	Ti $\longrightarrow$ Zn $3d^24s^2$ $3d^{10}4s^2$	Ga $3d^{10}4s^24p^1$	Ge $3d^{10}4s^24p^2$	As $3d^{10}4s^24p^3$	Se $3d^{10}4s^24p^4$	Br $3d^{10}4s^24p^5$	Kr $3d^{10}4s^24p^6$

Chapter 1
Structure and Bonding

1.1 Electronic structure

The electronic structures of the elements in the first three periods of the periodic table are shown in Table 1.

From its outer-shell electronic configuration of $3s^2 3p^3$ and a consideration of Hund's rule, the bonding electrons of phosphorus should be best represented by $3p_x^1 3p_y^1 3p_z^1$, which in turn should lead to the formation of tervalent compounds, as with nitrogen. It has been suggested, however, that phosphorus might be able to use its 3d orbitals in bonding by promoting electrons from the 3s level. As we shall see, a great deal of evidence has accumulated which makes a dismissal of this suggestion very difficult.

A study of the energies involved indicates that it will be difficult to promote an electron from the 3s to the 3d level in phosphorus, some 17 eV being required (which is equivalent to about 1780 kJ mol^{-1} (425 kcal mol^{-1}) – a large amount of energy to be available in any normal chemical reaction). In the case of nitrogen the promotional energy from 2s to 3d is even higher, about 23 eV or 2410 kJ mol^{-1} (576 kcal mol^{-1}) and many of the differences between the chemistry of nitrogen and phosphorus have been explained in terms of the greater difficulty experienced by nitrogen in using its 3d orbitals. The suggested high energy of the 3d level is supported by potassium and calcium, which start the third period, actually filling the 4s level before the 3d level is used. However, the importance of Hund's rule in controlling configuration in the transition metals shows that 4s and 3d are very nearly equivalent in energy. One of the few definite statements that can be made relating to discussions of this type is that phosphorus will not use its 3d orbitals unless the energy gain from their use is at least equivalent to the promotional energy involved.

Phosphorus, like carbon, silicon, nitrogen and other elements near the centre of the periodic table, shows a high ionization potential. This inhibits the formation of ions, and so the chemistry of phosphorus, at normal temperatures, is almost entirely that of covalent compounds.

1.2 Bonding in trivalent-phosphorus compounds

The bonding in trivalent phosphorus compounds can be discussed in terms of s-orbitals and p-orbitals (Figure 1). The basic principles involved in bonding of

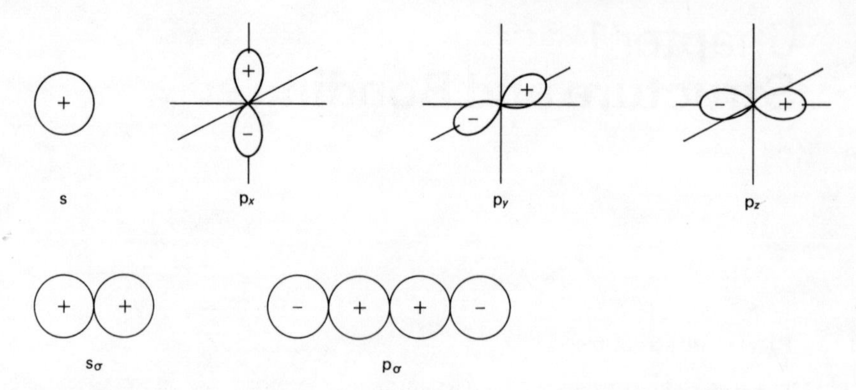

Figure 1

this type have been fully dealt with in a number of publications (see Bibliography) and so the discussion here is confined to the specific application of these principles to bonding in phosphorus compounds.

Figure 1 shows the radial distribution functions for 3s and 3p orbitals, and a representation of their use in forming σ-bonds. The symmetry requirements of these bonds are simple, and a graph of orbital overlap against nuclear separation is shown in Figure 2(a).

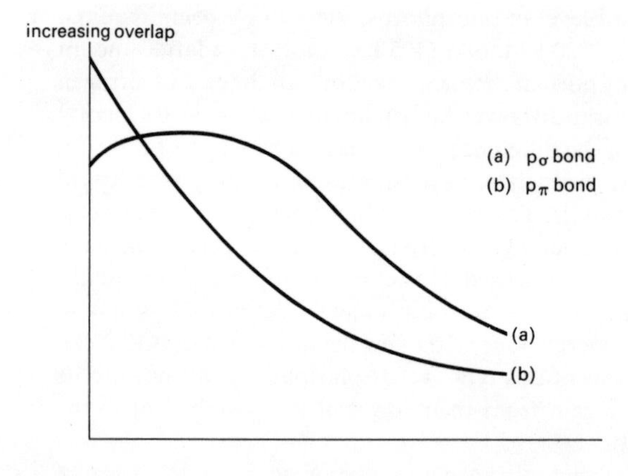

Figure 2

The p_x, p_y and p_z orbitals in the 3p level are mutually at right angles (orthogonal) and so σ-bonded trivalent phosphorus compounds would be expected to show bond angles close to 90°. Although this is not entirely true, it is more nearly the case for phosphorus than for nitrogen compounds, where bond angles are generally closer to the tetrahedral angle (108°). (See p. 22, Table 6.)

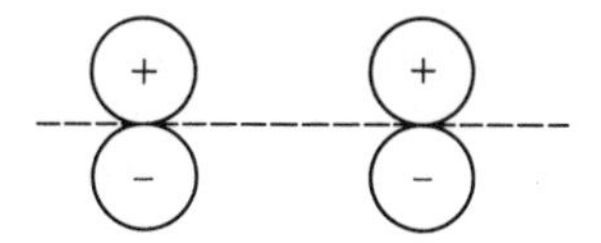

Figure 3

p-Orbitals can also form π-bonds by the overlap of those p-orbitals at right angles to a line between the nuclei, rather than those directed along it (Figure 3).

Orbital overlap as a function of nuclear separation for p_π–p_π bonding is shown in Figure 2(b). The difference in shape from 2(a) is expected from simple symmetry considerations; maximum overlap will not be reached in the case of p_π bonds until the internuclear separation is zero. An interesting conclusion which can be drawn from this graph is that the p_π overlap would be expected to be greater for the elements in the first row of the periodic table than for those in the second and later rows, because internuclear separations will increase on moving down each group.

1.2.1 *Bond strengths*

Pauling's 'bond energies' are calculated from enthalpy changes and always relate to the gaseous state. The resultant value is the sum of energies from all forms of bonding (polar, non-polar, σ and π). However, it is possible to estimate individual contributions from each form of bonding by considering compounds which only show one type of bonding. The energy contribution due to this form can be used in discussing the properties of other compounds in which this bond is present.

An approach of this type has been used by Pauling to develop an electronegativity scale (Table 2). A similar approach can be used to estimate the energies of π-bonds, and hence the contribution of π-bonding to any particular bond. A comparison of σ-bond and π-bond energies for the diatomic molecules of some elements near phosphorus in the periodic table is given in Table 3. Com-

Table 2 Pauling's electronegativity scale

H			
2·1			
C	N	O	F
2·5	3·0	3·5	4·0
Si	P	S	Cl
1·8	2·1	2·5	3·0
Ge	As	Se	Br
1·8	2·0	2·4	2·8

Pauling (1960), p. 93.

parison of these values shows that carbon has an abnormally high σ-bond energy, which is reflected in the stability of its compounds. In general, however, the homopolar σ-bonds formed by the second-row elements are stronger than the analogous bonds formed by the first-row elements. The situation becomes far less clear when heteropolar bonds are considered (Table 4), although some tentative conclusions can be drawn, for example, phosphorus–halogen bonds are much more stable than nitrogen–halogen bonds on purely energetic considerations, which can be confirmed by anyone who has attempted to prepare any of the nitrogen trihalides. On moving to the third-row elements, the σ-bond energies are seen to fall again, a factor which is important in the rate of formation of quaternary salts of the elements of group V

$R_4P^+X^- > R_4N^+X^- > R_4As^+X^-$.

Table 3 Homopolar bond energies/kJ mol^{-1} (kcal mol^{-1})

Type of Bond	Group IV	Group V	Group VI	Group VII
	C	N	O	F
σ	348 (83)	159 (39)	138 (35)	152 (36)
π	264 (63)	272 (65)	352 (84)	
	Si	P	S	Cl
σ	222 (53)	209 (50)	226 (54)	239 (57)
π		142 (34)	126 (30)	
	Ge	As	Se	Br
σ	159 (39)	134 (34)	184 (44)	193 (47)

Table 4 Heteropolar bond energies/kJ mol^{-1} (kcal mol^{-1})

Carbon			*Nitrogen*		*Oxygen*	
C—H	C—O	C—Cl	N—H	N—Cl	O—H	O—Cl
415	352	327	390	197	460	201
(99)	(85)	(78)	(93)	(46)	(110)	(48)
C—F				N—F		
440				270		
(105)				(65)		
Silicon			*Phosphorus*		*Sulphur*	
Si—H	Si—O	Si—Cl	P—H	P—Cl	S—H	S—Cl
293	369	356	318	331	339	251
(70)	(88)	(86)	(77)	(79)	(81)	(60)
			P—O	P—F		
			360	526		
			(86)	(126)		

The variation in π-bond energies is the antithesis of that in σ-bond energies. The elements of the second row show a large fall in π-bond energy over those of the first row. Possible reasons for this have already been discussed (p. 17).

The weaker π-bonds of the second-row elements are much in evidence in the chemistry of phosphorus. Examples of phosphorus compounds which show multiple bonding involving p-orbitals are rare. A number of compounds which were originally thought to possess multiple bonds have since been shown to be dimers, or higher polymers. For example, phosphorus imide–amide was thought to be HN=P—NH_2, but has since been shown to be a polymer. The structure of 'phosphobenzene' $(PhP)_n$ was thought to be PhP=PPh until some ten years ago when it was shown to be of complex composition, but mainly the cyclic compound (1.1) (see p. 237). The analogous nitrogen compound is the well-known azobenzene, Ph—N=N—Ph.

```
Ph\       /Ph
   P —— P
   |    |
   P —— P
Ph/       \Ph
```

(1.1)

Any comparison of compounds of nitrogen and phosphorus reveals the reluctance of phosphorus to form multiple bonds, for example, the totally different structures of the nitrites (R—O—N=O is a monomer) and their phosphorus analogues (attempts to prepare R—O—P=O lead to polymeric material 1.2).

```
\
 P—O [—P—O ]—
 |      |
 OR     OR  ]n
```

(1.2)

Substituted phosphorus analogues of pyridine (1.3) and pyrrole (1.4) have been prepared, although the chemical reactions of the latter – for example, participation in Diels–Alder reactions and strong donor properties – do not suggest an involvement of the phosphorus lone pair in the ring. Measurements of heats of hydrogenation however suggest that there is limited conjugation in the fluorene analogue (1.5). Apart from some isolated examples, which on further study seem likely to be polymeric, the only remaining structures which may constitute examples of stable p_π-bonded phosphorus compounds are the phosphacyanines, for which structure (1.6) has been suggested, and recently confirmed by X-ray analysis.

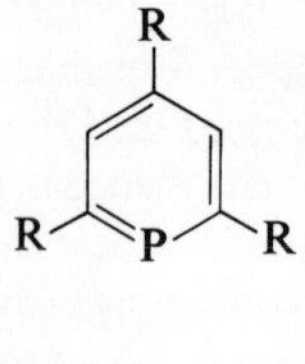

(1.3) 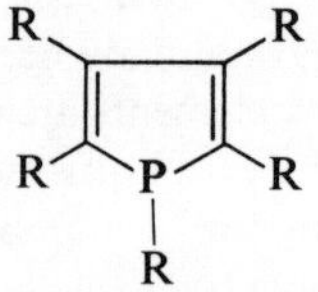 (1.4) 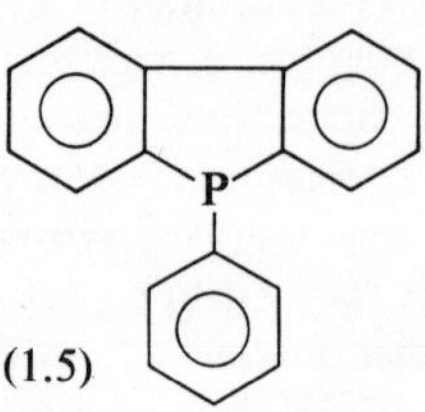 (1.5)

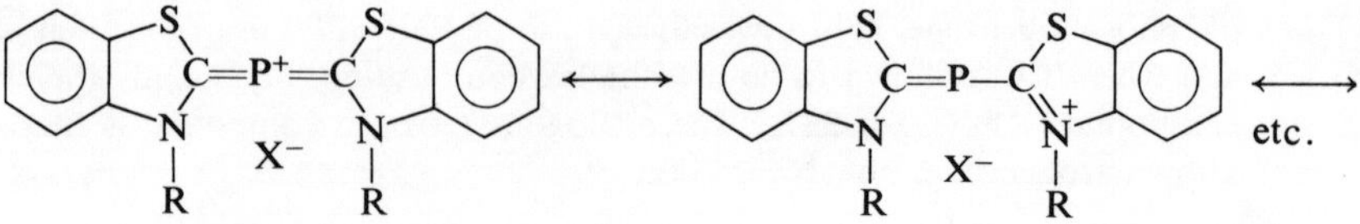

(1.6)

Hudson (1965) suggests that the preparation of compounds such as $R{-}P{=}C{<}^{R}_{R}$ and R—P=P—R will be possible if the groups R are capable of conjugation, and this type of stabilization may well be sufficient to stabilize the phosphacyanines.

Examples which illustrate the low π-bond energy are found in the chemistry of elements throughout the second row; thus silicon analogues of acetone, acetylene, carbon dioxide, etc. have yet to be isolated. Although a number of explanations of this phenomenon, including the much simplified one discussed earlier, have been postulated, the problem cannot yet be regarded as solved.

It is important to realize that one form of π-bonding may well have a wide existence in phosphorus compounds, that which involves d-orbitals, but it seems clear that p_π bonding plays very little part in the bonding of stable phosphorus compounds.

1.2.2 *Bond lengths*

The σ-bond length of a bond between two atoms is obtained by simple addition of the σ-bond radii of the atoms involved. These radii have been determined for each element from the bond length of a single σ-bond between two atoms of that element. Unfortunately, as is often the case in chemistry, the theoretically calculated σ-bond length between two atoms and the actual bond length, as determined experimentally by X-ray or electron diffraction, do not usually agree.

This lack of agreement is due to two factors. Firstly, the actual bond may have some π-character which will lead to a bond shortening, and secondly, unless the bond is homopolar, it will have some ionic character. The amount of ionic character should be directly dependent on the electronegativity difference and a rule has been developed (the Schomaker–Stevenson rule) which relates these values numerically, allowing a direct calculation of the ionic character in any bond from electronegativity values.

Table 5 Bond lengths and π-bond character

Compound	*Bond*	*Bond length*/10^{-11} m	*π-Bond* character*
3-coordinate phosphorus			
P_2	P—P	1·890	2·0
PH_3	P—H	1·424	0·1
PF_3	P—F	1·546	0·2
PCl_3	P—Cl	2·00	0·0
$P(CH_3)_3$	P—C	1·87	0·1
P_4 (crystal)	P—P	2·205	0·0
4-coordinate phosphorus			
F_3PO	P—O	1·56	0·4
	P—F	1·52	0·3
Cl_3PO	P—O	1·45	1·0
	P—Cl	1·99	0·0
F_3PS	P—S	1·85	1·0
	P—F	1·51	0·3
Cl_3PS	P—S	1·94	0·4
	P—Cl	2·01	0·0
5-coordinate phosphorus			
PF_5	$P—F_{equatorial}$	1·57	0·2
	$P—F_{apical}$	1·59	0·1
PCl_5	$P—Cl_{equatorial}$	2·01	0·0
	$P—Cl_{apical}$	2·07	0·0

* According to Schomaker and Stevenson (1941).

The π-character of the bond is equivalent to any bond shortening beyond that expected for that amount of ionic character. By calculations of this type Pauling has shown that one π-bond leads to a bond shortening of $0{\cdot}19 \times 10^{-11}$ m on average, whereas two π-bonds will shorten a pure σ-bond by $0{\cdot}30 \times 10^{-11}$ m. Table 5 shows a number of bond lengths together with the calculated amount of π-bonding character.

Table 5 indicates that, according to bond-length considerations, trivalent-phosphorus compounds show only very minor π-bonding. This supports the conclusion reached from a study of bond energies that π-bonding involving p-orbitals is very unlikely in the case of phosphorus. π-Bonding does appear to become much more important in four-coordinated phosphorus compounds, and is in fact one piece of evidence for the participation of d-orbitals in the bonding of these compounds (see pp. 32–6). The apparent support that conclusions derived from bond energies appear to gain from a study of bond lengths should not be overemphasized because, although it seems likely from this agreement that the same factor is influencing both values, that factor need not be π-bonding. However, it is a possibility which conveniently fits the known facts.

1.2.3 *Bond angles*

The bond angles exhibited by any particular compound are dependent upon the state of hybridization of the central atom. In Table 6 the bond angles of a number of trivalent-phosphorus and nitrogen compounds are listed. It is apparent from these values that nitrogen compounds generally have bond angles about 108° (the tetrahedral angle), whereas with phosphorus the values range from 93° for phosphine to about 100° for the phosphorus trihalides (although an earlier value for phosphorus trifluoride was 104°). It seems reasonable to conclude from these values that phosphorus uses mainly p-orbitals with a small amount of s-character (a conclusion drawn from the directional characteristics of p-orbitals) while in trivalent-nitrogen compounds the central nitrogen atom is mainly sp^3 hybridized. The anomalous value for the bond angles in P_4, which has structure (1.7), has been explained by more complex hybridization involving d-orbitals. However, the most recent theory involves the use of bent p-orbitals analogous to the bonding in cyclopropane, where the highest electron density is not on a direct line between the two nuclei.

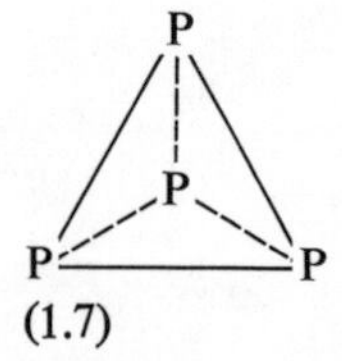

(1.7)

Table 6 Bond angles

Phosphorus compound	*Angle*	*Value*	*Nitrogen compound*	*Angle*	*Value*
P_4	P—P—P	60°			
PH_3	H—P—H	93°	NH_3	H—N—H	107°
			$MeNH_2$	C—N—H	112°
			Me_2NH	C—N—C	108°
Me_3P	C—P—C	100°	Me_3N	C—N—C	108°
$(CF_3)_3P$	C—P—C	99°	$(CF_3)_3N$	C—N—C	114°
F_3P	F—P—F	100°	F_3N	F—N—F	102°

The ability of phosphorus to show bond angles near 90° enables it to take part in the formation of small rings and there are numerous examples of stable phosphorus compounds involving phosphorus in four-membered rings, for example (1.8–11).

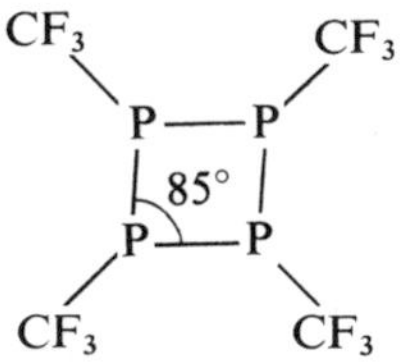

CF_3 CF_3
P—P
85°
P—P
CF_3 CF_3
(1.8)

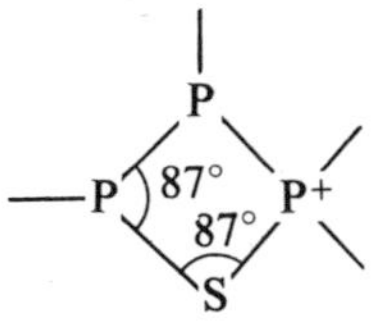

P
—P 87° P^+
87°
S
(1.9)

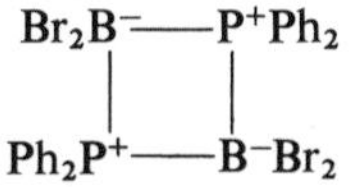

Br_2B^-—P^+Ph_2
Ph_2P^+—B^-Br_2
(1.10)

S P S P CH_3
CH_3 S S
(1.11)

The variation of bond angles in trivalent-phosphorus compounds is dependent on a number of factors. The bond angle might be expected to depend on the relative interatomic repulsions of the groups attached to the phosphorus. If these groups are very electronegative they would be expected to have a considerable negative charge residing on them. This would lead to a larger X—P—X bond angle (this is supported by the 100° angle for PF_3). However, it would be dangerous to draw too many firm conclusions from this type of discussion because other factors such as steric effects, the modification of hybridization of the phosphorus atom due to electronegative substituents and the possible intervention of d_π bonding also play a part in any trends observed.

1.2.4 *Dipole moments*

Hybridization, and hence bond angles, are also intimately involved with dipole moments. Part of the dipole moment of a compound can be attributed to the polarity of its bonds, but attempts to correlate these factors have been unsuccessful because dipole moments are in fact due to a summation of variables such as polarization, hybridization and π-bonding effects in addition to the ionic character of the bonds.

A comparison of the dipole moments of some tertiary nitrogen and phosphorus compounds (Table 7) shows an inverse dependence of dipole moment on substituents in amines and phosphines. Thus, although ammonia has a larger dipole moment than trimethylamine, the moment of phosphine is smaller than that of trimethylphosphine. A satisfactory explanation of this phenomenon is not at present available. However, it seems possible that the large difference in bond angle between phosphine and trimethylphosphine, compared with the negligible difference between ammonia and trimethylamine, may be important, as it is

Table 7 Dipole moments

Amines	*μ/Debyes* (μ/C m × 10^{-30})	*Phosphines*	*μ/Debyes* (μ/C m × 10^{-30})
NH_3	1·4 (4·7)	PH_3	0·6 (2·0)
$MeNH_2$	1·23 (4·10)		
Me_3N	0·6 (2·0)	Me_3P	1·8 (6·0)
Et_3N	0·82 (2·73)	Et_3P	1·84 (6·14)
Ph_3N	0·26 (0·87)	Ph_3P	1·44 (4·80)
		Me_2PPh	1·22 (4·07)

known that small changes in the hybridization of lone pairs of electrons can produce large changes in dipole moments. Other factors which will undoubtedly play a part are the greater electronegativity of nitrogen over phosphorus and the difference in quantum number of the orbitals used in each lone pair.

Attempts have been made to use dipole-moment measurements to determine the extent of p_π bonding in some phosphorus compounds. The dipole in trimethylphosphine is shown in (1.12). It seems reasonable that if any p_π bonding is present in dimethylphenylphosphine (one possible canonical form of this is shown in 1.13) it will lead to a dipole in opposition to that shown in trimethylphosphine. Thus the difference in dipole moment between trimethylphosphine and dimethylphenylphosphine should give a measure of the p_π bonding present in the latter. It would be dangerous to place too much reliance on determinations of this type, because, as we have already seen (p. 23), dipole moments are dependent on various factors.

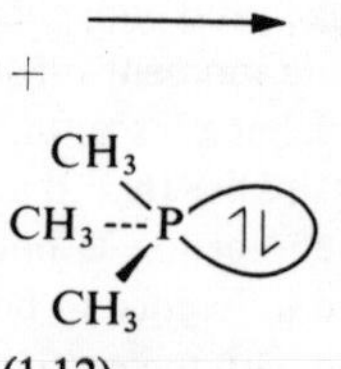

CH_3, CH_3---P, CH_3 (with lone pair and dipole arrow $+\!\!\longrightarrow$)

(1.12)

CH_3, CH_3 >P^+= (cyclohexadienylidene ring) $^-$

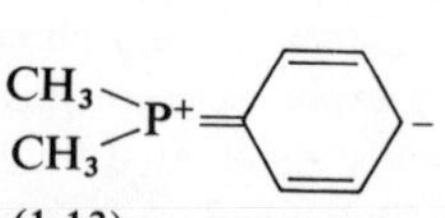

(1.13)

1.2.5 *Basicity and nucleophilicity*

The abilities of a compound to act as a base and as a nucleophile are generally separated by chemists, because these properties may be very different for any particular compound. This is due to the unusual properties of the proton (the ease of the addition of which controls basicity) when compared to other electrophilic centres. Trivalent-phosphorus compounds, although generally good nucleophiles, are very weak bases (see Table 8). The variation in pK_a is much greater for the phosphines than for the amines. Although the pK_a values of trimethylamine and trimethylphosphine are comparable, that of phosphine is very much less than that of ammonia. These pK_a values will depend on a number of factors, not least of which is the availability of the lone pair on the central atom. From the earlier discussion it is apparent that trivalent-phosphorus compounds tend to show more p-character in their bonds, and hence more s-character in their lone pair, than analogous nitrogen compounds. The large s-character of the lone pair causes it to spend much more of its time in the vicinity of the nucleus, as evidenced by the distribution functions of s-orbitals and p-orbitals (Figure 1).

This difference in hybridization is important in another way. After protonation the resulting phosphonium, or ammonium, cation will be sp^3 hybridized in bonding to its four substituents. While this requires little or no change in hybridization for amines, considerable rehybridization is needed in the case of phosphines. This would be expected to decrease pK_a values.

Table 8 pK_a values for amines and phosphines

Phosphine	pK_a	*Amine*	pK_a
PH_3	−14	NH_3	9·21
$MePH_2$	−3·2	$MeNH_2$	10·62
i$BuPH_2$	−0·02	*i*$BuNH_2$	10·52
Me_2PH	3·9	Me_2NH	10·64
(*i*Bu)$_2$PH	4·1	(*i*Bu)$_2$NH	10·50
Me_3P	8·65	Me_3N	9·76
(*i*Bu)$_3$P	7·97	(*n*Bu)$_3$N	10·89

The apparently abnormal variation in going from dimethylamine to trimethylamine is due to steric effects; that it is not observed with phosphines is due to the increased atomic radius. An alternative explanation of the differences in pK_a values between nitrogen and phosphorus compounds is available from heats of solvation and it is apparent that these do indeed account for part, if not all, of the difference.

Tertiary phosphorus compounds act as excellent nucleophiles towards carbon, most examples being concerned with attack at saturated carbon, for example, formation of phosphonium salts with alkyl halides (p. 53). Nucleophilic attack at phosphorus also takes place, for example

$$(EtO)_3P{=}O + H_2O \rightarrow (HO)_3P{=}O.$$

However, it appears to be controlled by a different scale of nucleophilicity to that of saturated carbon. A great deal of further work will be necessary before any firm conclusions can be drawn, but initial impressions are that the main differences are related to bond strengths in the products, for example, F^- is a good nucleophile towards phosphorus, but poor towards carbon (cf. C—F and P—F bond strengths in Table 4).

1.2.6 *Inversion and optical activity in phosphines*

The energetics of the inversion process are shown in Figure 4, which shows the relationship between the potential energy of the system and the distance x of the groups R^1, R^2 and R^3 from a point A directly below the central atom.

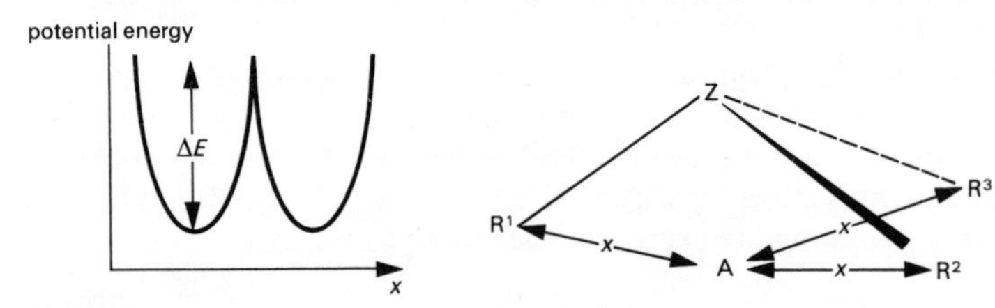

Figure 4

The value of ΔE (the energy barrier to inversion) for any particular compound will control its rate of inversion. Some values of ΔE are given in Table 9.

A phosphine molecule (1.14a) is asymmetric, and hence will give rise to optical activity as long as its rate of inversion (1.14a → b) is sufficiently slow.

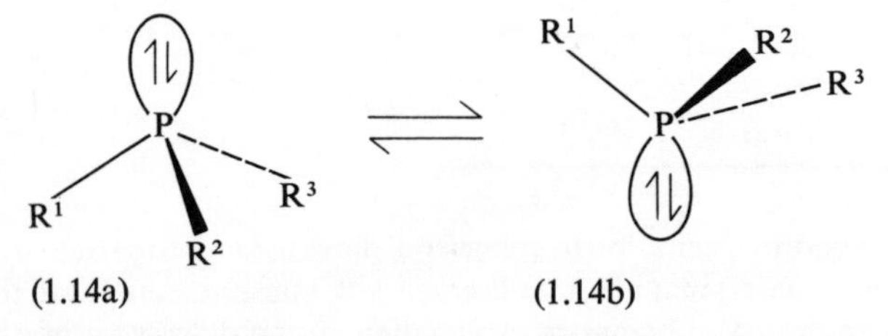

A reasonable value of ΔE which would allow resolution of the two forms at room temperature has been calculated to be about 105 kJ mol^{-1} (25 kcal mol^{-1}). From this we would expect that phosphines would probably be resolvable, but most amines not. Arsines should be readily resolvable even at high temperatures and it has been calculated that the half-life for inversion of AsH_3 is 1·4 years at room temperature, while that of AsD_3 is $3{\cdot}5 \times 10^7$ years, showing the extreme dependence of the rate of inversion on the mass of substituents.

The first example of optical activity attributed to an asymmetric trivalent group V atom involved an arsenic compound. However, in more recent years a number of examples of optically active phosphines have been prepared by a variety of methods. Probably the most useful of these is the electrolytic reduction of phosphonium salts (p. 130) discovered by Horner. Phosphonium salts are fairly simple to resolve and their reduction in this manner proceeds with retention of configuration.

$$R^1R^2R^3P^+{-}CH_2{-}Ph + 2e \xrightarrow{2H^+} R^1R^2R^3P{:} + PhCH_3$$

More recently Mislow has developed a method of preparing optically active phosphine oxides by resolution of menthyl phosphinates (1.15). These react

Table 9 Energy barriers to inversion

Amines	ΔE/ (kJ mol^{-1})	*Phosphines*	ΔE/ (kJ mol^{-1})
NH_3	24·8	PH_3	115
	(25·1)		(72·7)
ND_3	23·7	$PhCH_3P{-}PCH_3Ph$	109†
Me_3N	34·2	Me_3P	133

Arsines	ΔE/ (kJ mol^{-1})	*Stibines*	ΔE/ (kJ mol^{-1})
AsH_3	146		
	(134)		
$(CH_3)_3As$	122	Me_3Sb	112

Weston (1954), p. 2645; values in parentheses from Costain and Sutherland (1952).
† Lambert, Jackson and Mueller (1968), p. 6403.

with Grignard reagents with retention of configuration to give phosphine oxides, for which specific methods of reduction to phosphines are available with either retention or inversion of configuration.

(1.15) → $R^1R^2R^3P{=}O$ phosphine oxide —retention→ $R^1R^2R^3P{:}$ phosphine; —inversion→ $R^1R^3R^2P{:}$

Diphosphines (1.16) possess two adjacent asymmetric centres and show optical properties analogous to those of tartaric acid ((+), (–) and meso forms). Similar behaviour has been described for diphosphine dioxides (1.17) and disulphides (1.18).

$R^1R^2P—PR^1R^2$
(1.16)

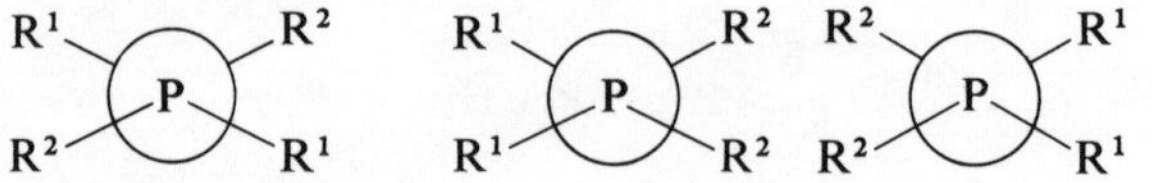

meso form

(+) and (–) forms

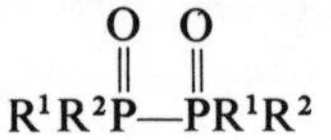

$R^1R^2P(=O)—P(=O)R^1R^2$
(1.17)

$R^1R^2P(=S)—P(=S)R^1R^2$
(1.18)

As we shall see later (p. 130), optically active phosphines, phosphine oxides and phosphonium salts are invaluable for determining mechanisms in phosphorus chemistry.

1.3 The use of d-orbitals in phosphorus chemistry

Undoubtedly the most controversial feature of phosphorus chemistry in recent years has been the use of d-orbitals in bonding. Before any attempt can be made to assess the evidence available in support of this type of bonding, it will be necessary to become familiar with the d-orbitals themselves. What are their basic shapes and symmetries? What types of bonds might they be expected to form?

1.3.1 *The* d-*orbitals*

The 3d level contains five d-orbitals (their probability distributions and symmetries are shown in Figure 5). Each of these is capable of containing a maximum of two electrons of opposite spin.

It is evident from Figure 5 that symmetry considerations in bonding involving d-orbitals are going to be much more complex than those for p-orbitals and s-orbitals. This, together with the large energy difference involved, might be expected to reduce the chance of d-orbitals taking part in hybridization with s-orbitals and p-orbitals.

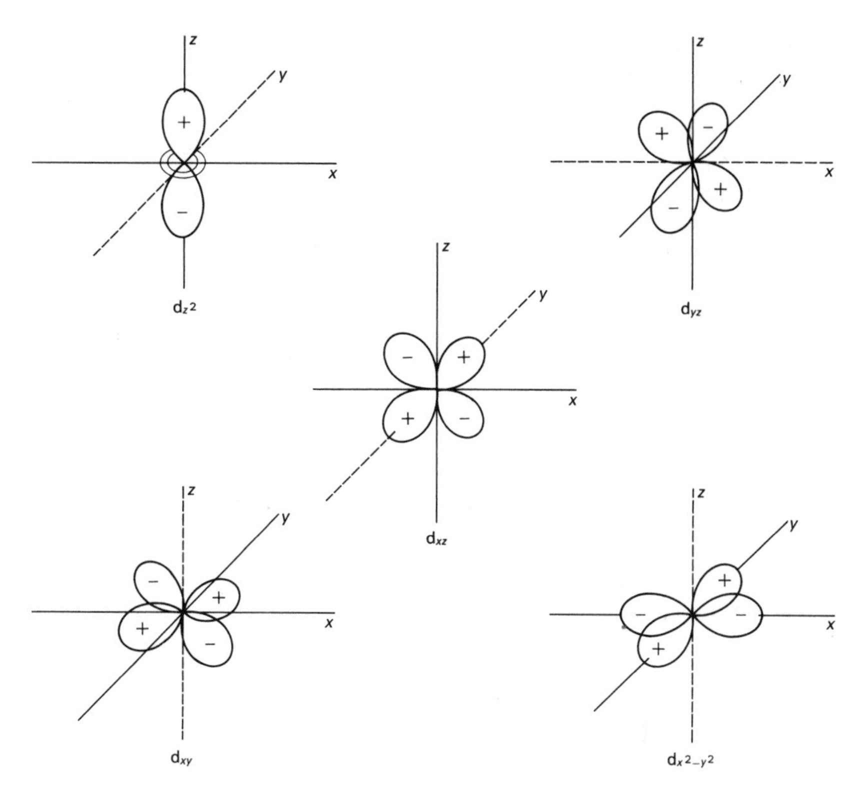

Figure 5

.3.2 d_σ *Bonding*

Unlike d_π bonding, the existence of d_σ bonding in phosphorus compounds is well established. In any compounds where the phosphorus has more than four groups attached to it, some orbital of higher energy than the 3p must be used in bonding unless the compound is ionic. Although the 3d orbitals are much higher in energy than the 3p (p. 15), they are the lowest-energy orbitals available with the symmetry needed to explain the known structures of five-coordinate phosphorus compounds.

Pauling originally suggested that all phosphorus compounds of coordination number greater than four are ionic. On the basis of the evidence available at that time this was quite reasonable and obviated the need to postulate the use of 3d orbitals. Indeed phosphorus pentachloride and pentabromide have the structures $(P^+Cl_4)(P^-Cl_6)$ and $(P^+Br_4)(Br^-)$ in the solid state (although of course the six-coordinate anion present in phosphorus pentachloride does require the use of d-orbitals). More recently, however, the formerly small number of five- and six-coordinated phosphorus compounds known has rapidly expanded. Many of these have been shown to be covalent and to have stereochemistries which are not compatible with ionic structures.

Two factors which must be kept in mind in any consideration of bonding are that the energies and radial distribution functions (i.e. the physical significance in space) of the orbitals involved must be comparable and that the angular distribution function (or symmetries) of the orbitals must be such that a bond, and not an antibond, is formed; see Figure 6. As we shall see (p. 32), the factors involved in the second of these criteria are much more complex for d_{π} bonding than for d_{σ} bonding.

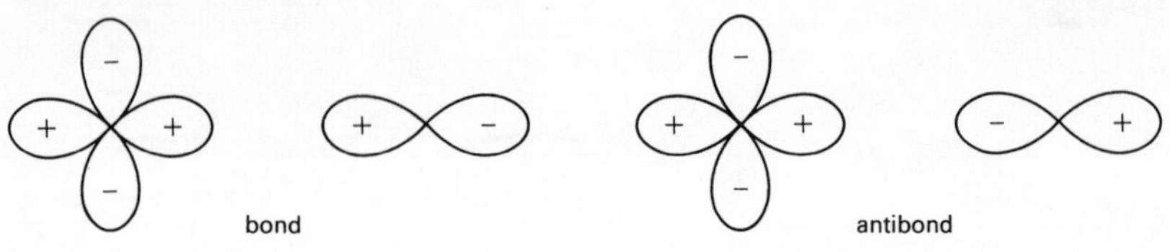

Figure 6

From considerations of this type three different hybridizations of s-, p- and d-electrons for five-coordinate phosphorus have been postulated, leading to the stereochemistries shown in Figure 7.

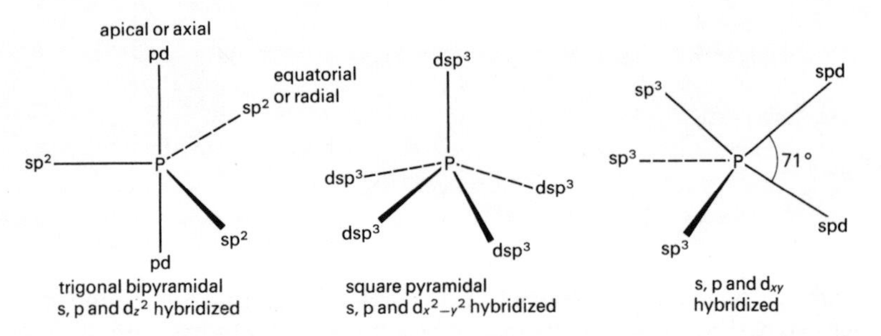

Figure 7

Each of these hybridized states is made up of one s-orbital, three p-orbitals and one d-orbital (thus five original orbitals lead to five hybrid orbitals), the stereochemistry of each being determined by the symmetry of the particular d-orbital involved in the hybridization.

The most stable state for the square-pyramidal structure is when the phosphorus atom is slightly above the plane which passes through the four basal ligands, although even then the trigonal-bipyramidal structure has a slightly lower energy. The third possible structure, involving sp^3 and spd hybrids, has the highest energy of all.

Most pentavalent-phosphorus compounds so far studied appear to have the trigonal-bipyramidal structure (e.g. phosphorus pentachloride, liquid, gas and solution; phosphorus pentafluoride; pentaphenylphosphorus). In fact, no example of a square-pyramidal phosphorus compound has yet been found (although pentaphenylantimony is square pyramidal in the solid). Calculations indicate that for the trigonal-bipyramidal structure the two apical bonds should be longer than the three radial bonds. Support for this can often be obtained by experiment, for example, fluorine-19 nuclear magnetic resonance spectroscopic studies have shown that there are two types of fluorine present (electronegative elements tend to be apical, see p. 111).

Originally it was thought that the 3d orbitals of a neutral atom would be too diffuse for meaningful overlap with the 2p orbitals of the first-row elements. In the light of this, compounds in which the phosphorus atom carried a positive charge would be expected to make more use of d-orbitals in their bonding, because electrostatic attraction will increase the d-orbital density, for example, the phosphorus ylids (p. 138) appear to show d_{π}–p_{π} bonding. Some support for this suggestion is available in that electron-withdrawing substituents on phosphorus do appear to increase the stability of phosphorus(V) compounds, presumably by increasing the positive charge on phosphorus.

Further evidence in favour of this theory is found in the halides of sulphur. Sulphur tetrafluoride and sulphur hexafluoride are known but the equivalent compounds of chlorine or bromine are not. Sulphur, being more electronegative than phosphorus, might be expected to make better use of its d-orbitals anyway. Some doubt has been cast on discussions of this type in the case of sulphur by calculations which show that the 3d orbitals are far less diffuse than was originally thought. However, similar calculations have shown that the 3d orbitals are much more diffuse in the case of phosphorus.

Trigonal-bipyramidal structures can interconvert by a process known as *pseudo-rotation*. The importance of this type of process has only recently been realized. Consider a trigonal bipyramid; if we take any two equatorial substituents, and pull them out so that they both became apical, we will have converted one trigonal bipyramid into another. This process is known as pseudo-rotation and takes place in pentavalent-phosphorus compounds at a rate dependent on the substituents A, B, C, D and E. In the scheme below the bonds which became apical in each conversion are written by the equilibrium sign. Each new trigonal bipyramid can of course pseudo-rotate to three others.

D A / C—P—B / E ⇌(CB) A / D, C, P—B / E ⇌(DB) A C / D—P—B / E

⇅ DC

B A / D—P—C / E

This process can be a complication when optically active phosphorus compounds are being used in mechanistic studies which involve phosphorus(V) intermediates. A series of pseudo-rotations can lead to racemization at phosphorus, and care is required in the interpretation of such results.

The few known examples of six-coordinate phosphorus have octahedral configurations, for example, PCl_6^-, PF_6^-.

1.3.3 d_π *Bonding*

The original suggestions of d_π bonding were made when it became necessary to explain the very obvious differences between the chemistries of the first-row and second-row elements in the periodic table. (We have already seen how the use of d-orbitals to form σ-bonds explains the increase in maximum coordination number.)

The more complex symmetry requirements of d_π bonding over p_π bonding have already been pointed out and examples are given in Figures 8 and 9.

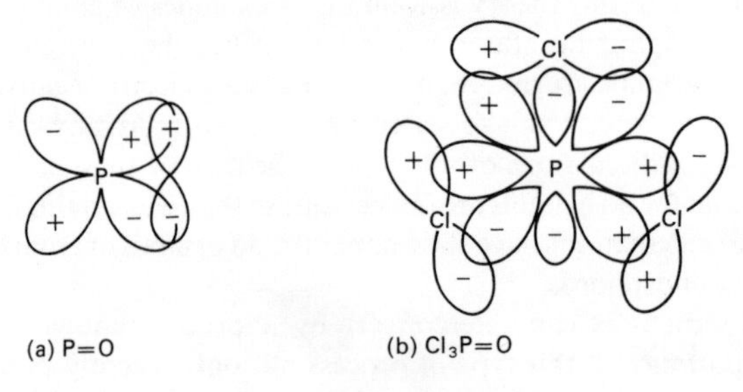

(a) P=O (b) $Cl_3P{=}O$

Figure 8

In the case of phosphoryl chloride (Cl_3PO) only one particular type of d_π–p_π bonding is shown, the chlorine atoms are, of course, also bonded to the central phosphorus by σ-bonds. In an attempt to simplify the figure the P=O bond has also been omitted.

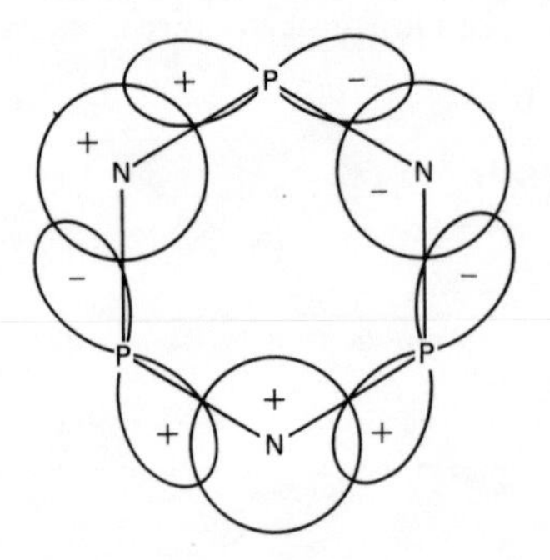

Figure 9

In the case of the P═O bond (the double bond representation is used for convenience) the d_π–p_π bonding is basically similar to the normal p_π–p_π case, but this is obviously not true in the other examples. It should also be noted that the p-orbitals involved in phosphoryl chloride have the same principal quantum number as the d-orbitals (3p from chlorine, 3d from phosphorus), whereas this is not true in the P═O bond (the lone pairs in oxygen being in 2p orbitals). From the symmetries of the orbitals involved it is clear that conjugation will not be complete in the case of the phosphonitrilic chloride trimer (Figure 9) because delocalization can only take place over a maximum of three atoms. This may well preclude the cyclic phosphonitrilics from showing any aromatic character.

In the case of phosphoryl chloride the chlorine atoms, together with phosphorus, are all involved in one molecular orbital and should show conjugation. However, in the P═O bond any d_π bonding is carried out by another pair of 3d orbitals and so we should not expect any conjugation of this with the chlorine substituents, even if d_π bonding does take place. This illustrates two important factors about d_π bonding. Firstly, that this bonding can take place independently to two, or more, groups on phosphorus and yet, unlike p_π–p_π bonding, these groups need not be in conjugation. Secondly, the symmetry of the d-orbitals is such that conjugation can take place without the criterion of planarity which is present in p_π–p_π bonding. It is due to the first of these factors that ultraviolet spectroscopy (which reveals conjugation) has been of little use, except in early misinterpretation, in determining the existence of d_π–p_π bonding. Considerations of this kind suggest that the stereochemical consequences of reactions as developed for carbon compounds may not be applicable to analogous reactions of phosphorus when d-orbitals are involved in its bonding.

One area of chemistry where phosphorus appears to use its 3d orbitals extensively in π-bonding is that of transition-metal complexes involving phosphorus ligands. Many of the original suggestions of d_π bonding came from studies in this field and until recently it was considered the only area of phosphorus chemistry in which d_π bonding could be postulated without much fear of contradiction. However, even here criticism has been voiced recently on the grounds that d_π bonding has been used to explain phenomena which can be readily explained without invoking d-orbital participation.

(a) *Physical evidence for* d_π *bonding*. (i) Bond parameters. The use of bond-energy measurements in the assessment of p_π–p_π bonding in phosphorus compounds has already been discussed and a similar approach can be made in an attempt to determine the extent of d_π–p_π bonding. In Table 10 bond energies for a variety of oxides and sulphides are shown, and it is apparent that the energy of the P═O bond is very much greater than the apparently analogous amine oxide N^+—O^- bond. It is suggested that this is due to the ability of phosphorus to use its 3d orbitals to reaccept some of the negative charge which it has donated to oxygen by way of its coordinate σ-bond (this is represented by resonance in 1.19). Nitrogen is unable to do this and so amine oxides only have a dipolar single

Table 10 Bond energies

Compound	$D_{P=X}$/ kJ mol^{-1} (kcal mol^{-1})	$D_{N^+-O^-}$/ kJ mol^{-1} (kcal mol^{-1})
	$D_{P=O}$	†
F_3PO	543 (129)	210–90 (50–69)
Cl_3PO	535 (128)	
Br_3PO	523 (125)	
Me_3PO	581 (139)	
Bu_3PO	573 (137)	
Ph_3PO	535 (128)	
$(EtO)_3PO$	631 (151)	
$(Me_2N)_3PO$	581 (139)	
	$D_{P=S}$	
$Pr_3P{=}S$	384 (92)	
$Bu_3P{=}S$	383 (92)	
Cl_3PS	290 (69)	
$(EtO)_3PS$	379 (91)	

Hartley, Holmes, Jacques, Mole and McCoubrey (1963).
† Hudson (1965).

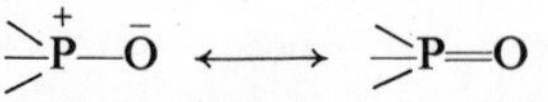

$$\gt\overset{+}{P}-\overset{-}{O} \longleftrightarrow \gt P{=}O \qquad \gt\overset{+}{N}-\overset{-}{O}$$

(1.19) 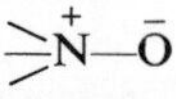 (1.20)

bond (1.20). The P=O bond is sometimes represented by P≡O because, as we have already seen (Figure 8a), there is the possibility of forming two d_π–p_π bonds, one with each oxygen lone pair.

Further support for this is available from a study of bond lengths. Whereas the N^+—O^- bond length in amine oxides is approximately the same as in hydroxylamine, the P=O bond length in phosphine oxides is $2{\cdot}5 \times 10^{-12}$ m less than that in the phosphites ($(RO)_3P$), again suggesting multiple bonding in the case of phosphorus, but not for nitrogen.

Similar arguments can be used in a discussion of the P=S bond in phosphine sulphides. However, it seems likely from a comparison of bond lengths and energies that there is less d_π–p_π bonding in P=S than in P=O. It is interesting that the P=O bond length in P_4O_{10} (1.21) is only $1{\cdot}39 \times 10^{-11}$ m, which is even less than the calculated value for P≡O.

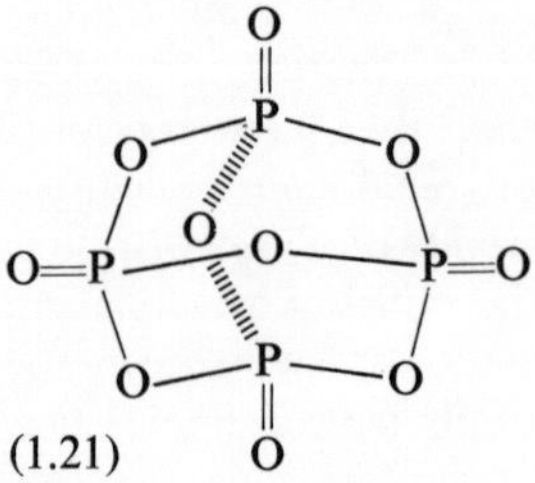

(1.21)

Bond-energy determinations in the phosphonitrilic chlorides (1.22) suggest that there is no great increase in stability over normal P—N bonds in acyclic compounds. But studies of bond lengths show that all the P—N distances around the ring are the same and equal to $1{\cdot}65 \times 10^{-11}$ m as compared with the calculated values of $1{\cdot}60 \times 10^{-11}$ m for P=N and $1{\cdot}80 \times 10^{-11}$ m for P—N. The P—Cl bond lengths also appear to be shortened. This can all be explained by d_π bonding.

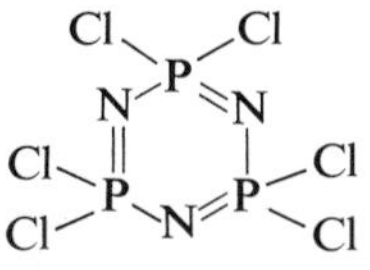

(1.22)

(ii) Dipole moments and pK_a values. Purely on the basis of electronegativity, the dipole moment due to R_3P^+—O^- should be greater than that due to R_3N^+—O^-, but the opposite is the case, and this has been suggested as evidence for d_π–p_π bonding in the phosphine oxides.

Further evidence is available from a study of donor ability in trivalent-phosphorus compounds. Where no π-bonding is possible, donor strengths follow expected σ-bonding ability, for example, $R_3P > (RO)_3P > Cl_3P > F_3P$. However, the same ligands show the reverse order of donor ability when coordinated to a central atom which possesses filled d-orbitals (e.g. a transition metal). It has been suggested that the central atom is able to back donate its excess charge via its d-orbitals into the empty d-orbitals of phosphorus (d_π–d_π bonding). This back donation would be enhanced by strongly electron-withdrawing substituents on the phosphorus atom, hence the reversal of ligand-bonding ability because the important factor is now acceptor rather than donor ability.

Studies of the pK_a values of the thio-acid (1.23) show that they are independent

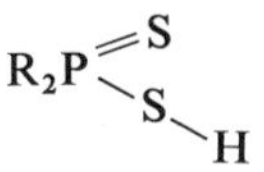

(1.23)

of R (this is evidence against conjugation through phosphorus rather than against d_π bonding). It is significant that in the sulphur compound (1.24) the hydrogen at the ring function is highly acidic, whereas that in the analogous carbonyl

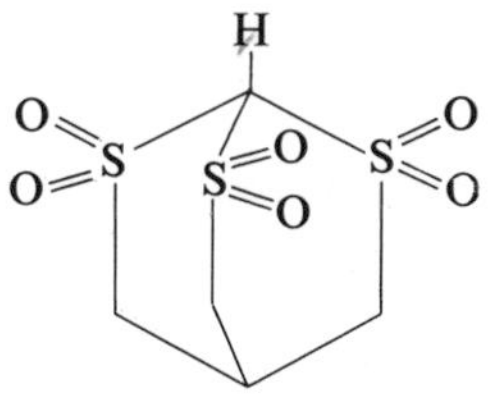

(1.24) 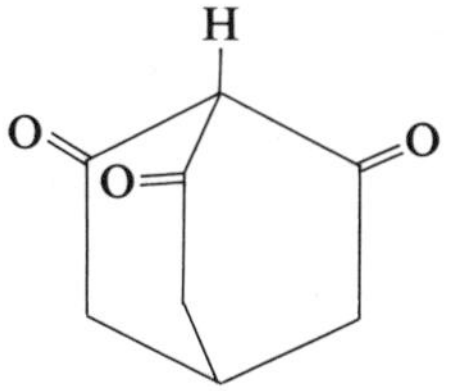 (1.25)

compound (1.25) shows very low acidity. This suggests that the angular requirements for d_π–p_π overlap, unlike those for p_π–p_π overlap, are not rigid.

Measurements of pK_a values in quaternary salts also produce evidence for d_π bonding. The protons adjacent to phosphorus in a phosphonium salt (1.26) are always more acidic than those in the analogous ammonium salt (1.27), contrary to what would be expected purely on electronegativity considerations.

$$\underset{(1.26)}{\overset{X^-}{R_3P^+\text{—}CH_2\text{—}R}} \qquad \underset{(1.27)}{\overset{X^-}{R_3N^+\text{—}CH_2\text{—}R}}$$

Systems of this type have been extensively studied because of their synthetic importance and this has led to considerable support for the postulate of d_π bonding.

(b) *Spectroscopic evidence for* d_π *bonding*. Early spectroscopic studies into d_π bonding were mainly concerned with ultraviolet spectra because of their success in the investigation of conjugation in p_π bonded systems. However, symmetry factors in d-orbitals may not allow conjugation through phosphorus (p. 33), although d_π bonding may still be taking place in various directions.

Even in the face of these drawbacks ultraviolet spectroscopy has still provided some useful evidence. The intense ultraviolet absorptions of cyclic polyphosphines (1.28) suggest electron delocalization between adjacent phosphorus atoms, a suggestion which is supported by Raman and infrared spectra. The ultraviolet spectra of phosphine sulphides $R_3P{=}S$ do not appear to be dependent on the polarity of the solvent. This has led to the suggestion that the P—S bond is zwitterionic in both the ground and excited states, and this has been used as evidence for the lack of d_π bonding in phosphine sulphides.

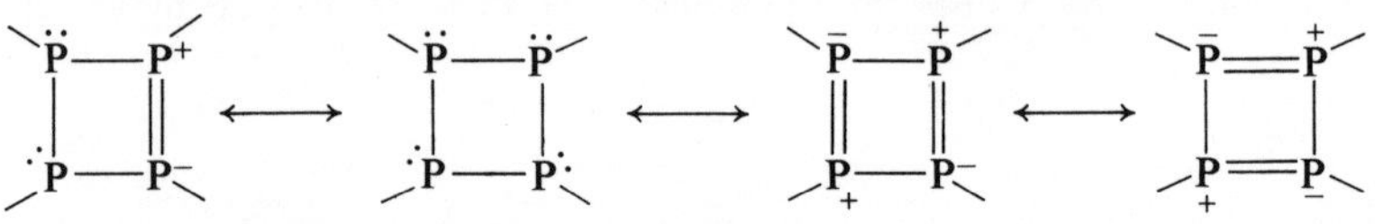

(1.28)

Changes in the infrared absorption frequencies of carbonyl ligands in transition-metal complexes have been correlated with the apparent d_π-bonding ability of other ligands on the same metal. As the substituent ligands increase their ability to accept electrons via their d-orbitals, the carbonyl stretching frequencies fall because they are forced to give up more electrons to the metal. The variation of P=O stretching frequency ($\nu_{P=O}$) with substituents on phosphorus has been studied, and a close correlation between the electronegativity of these substituents and $\nu_{P=O}$ has been found. Electronegative substituents would be expected to increase the stability of the P=O d_π–p_π bond but decrease the strength of the coordinate bond. Thus, a net increase in bond strength, and hence a rise in stretching frequency, with increasing electronegativity of substituents would support d_π bonding. This is found in the following order of P═O stretching frequency:

$PO_4^{3-} < MePO_3^{2-} < Me_2PO_2^- < Me_3PO$.

Infrared spectra also suggest an involvement of the lone pair of nitrogen in the ring of the cyclic compound (1.29). There is also some evidence for d_π bonding

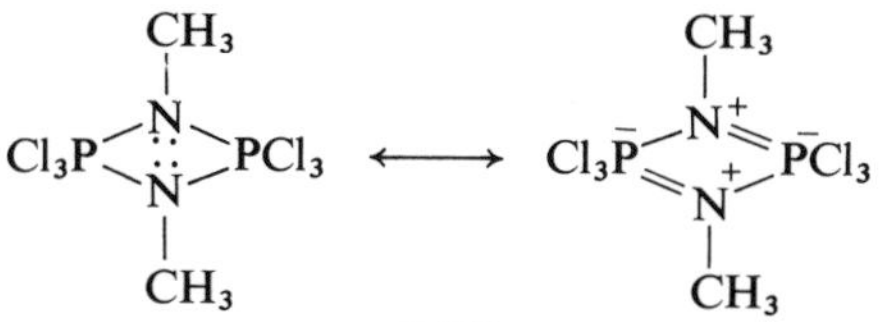

$$Cl_3P\langle(N\text{-}CH_3)_2\rangle PCl_3 \longleftrightarrow Cl_3\bar{P}\langle(\overset{+}{N}\text{-}CH_3)_2\rangle \bar{P}Cl_3$$

(1.29)

from chemical shifts in phosphorus-31 nuclear magnetic resonance spectra (see p. 41). More recently both nuclear quadrupole resonance and the Mössbauer effect have been used in the study of d_π bonding.

(c) *Chemical evidence for* d_π *bonding*. Evidence of this type can be found throughout organophosphorus chemistry. Some idea of the probable importance of this type of bonding can be gauged from the fact that the existence, and hence the chemistry, of phosphorus ylids, phosphonate anions, iminophosphoranes and many other synthetically important phosphorus compounds is best interpreted in terms of d_π bonding. The stability of phosphorus acids and esters and hence insecticides, nerve gases, phospholipids and nucleic acids is related by modern theory to d-orbital bonding in phosphorus.

The majority of chemical evidence is based on observations stemming from the high stability of ⋛P═O, ⋛P═C⋜ and ⋛P═N— bonds when compared with those of analogous nitrogen compounds, the energetics of which have already been discussed (pp. 33–6).

(i) P═X bond strength. Chemical evidence for the strength of the P═O bond is available from a myriad of sources. For example, secondary phosphites (1.30) are in tautomeric equilibrium with a phosphoryl form (1.31), an equilibrium which favours (1.31) to more than 95 per cent in almost all cases. The frequency with which the P═O bond is formed in reactions of phosphorus compounds lends further support to bond-energy measurements.

$$(RO)_2P\text{—}O\text{—}H \rightleftharpoons (RO)_2P(\text{═}O)H$$

(1.30) (1.31)

The phosphorus ylids are analogous to the phosphine oxides with the oxygen replaced by methylene and its homologues. Again these compounds are more stable than their nitrogen analogues, which are usually only prepared with difficulty at very low temperatures unless stabilized by strongly electron-withdrawing substituents.

There is considerable evidence that trialkylphosphorus ylids (1.32) are less stable than triarylphosphorus ylids (1.33). This could be due to the greater

$$R_3P{=}C{<}^{R^1}_{R^2} \qquad Ar_3P{=}C{<}^{R^1}_{R^2}$$

(1.32) (1.33)

conjugative ability of the aromatic group; such an explanation would support some form of delocalization across phosphorus between the aryl substituents and the ylid carbanion. However, an alternative explanation is that the aryl groups leave phosphorus with a larger positive charge than the alkyl groups and so improve the overlap potentialities of its d-orbitals in agreement with Slater's prediction (p. 31). If the bonding in phosphorus ylids was purely coordinate (had no π-contribution), any electron withdrawal from phosphorus would be expected to destabilize the ylid.

The different behaviour of quaternary ammonium and phosphonium hydroxides on pyrolysis can be explained by the availability of d-orbitals.

$$\underset{OH^-}{R_3P^+{-}CH_2{-}CH_2{-}R^1} \xrightarrow{\Delta} R_3P{=}O + R^1{-}CH_2{-}CH_3$$

phosphonium

$$\underset{OH^-}{R_3N^+{-}CH_2{-}CH_2{-}R^1} \xrightarrow{\Delta} R_3N{:} + R^1{-}CH{=}CH_2 + H_2O$$

ammonium

In the case of the phosphonium compound, the formation of a five-coordinated intermediate (1.34) is made possible by using the phosphorus 3d orbitals, while in the ammonium salt case these d-orbitals are not available.

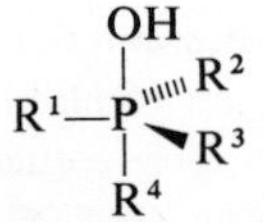

OH, R^2, $R^1{-}P$, R^3, R^4

(1.34)

(ii) Activation of adjacent multiple bonds. Unless activated by electron-withdrawing substituents, olefins are notoriously poor electrophiles (usually attributed to their peripheral π-electrons repelling attacking nucleophiles). However, vinylphosphonium salts (1.35) are readily attacked by a variety of nucleophiles. This presumes a considerable activation of the double bond by the phosphorus, an activation which could be due to d_π bonding stabilizing the ylid formed (1.36).

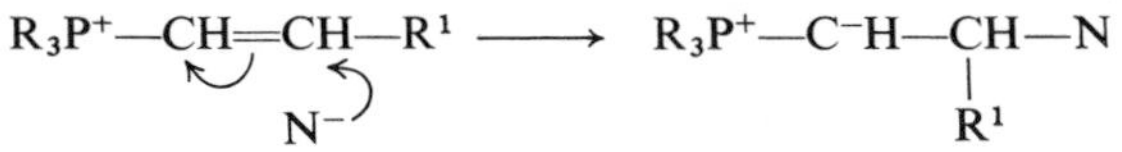

$$R_3P^+—CH{=}CH—R^1 \longrightarrow R_3P^+—C^-H—CH(R^1)—N$$

N^-

(1.35) (1.36)

That this is not a purely inductive effect is evidenced by the inert behaviour of the analogous vinylammonium salts towards nucleophiles (any inductive effect would be expected to be greater for nitrogen than for phosphorus on the basis of bond lengths). Similar reactivity is shown by vinylphosphine oxides, although here any d-orbital participation will be more complicated (although still possible), because presumably two d-orbitals are already involved in the P=O bond. The vinylphosphine oxide (1.37) is exceptional in that it is not attacked by nucleophiles. In fact it shows the nucleophilic character expected of unactivated olefins. This is thought to be due to the preferential interaction of the phosphorus d-orbitals with the nitrogen lone pairs rather than with the olefinic molecular orbital.

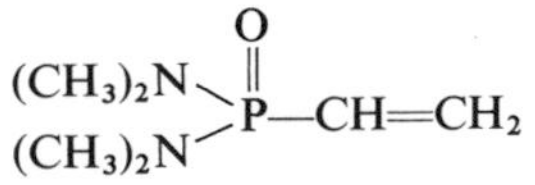

$$[(CH_3)_2N]_2P(=O)—CH{=}CH_2$$

(1.37)

(iii) Aromaticity. The study of aromaticity in phosphorus heterocycles has provided some evidence for d-orbital participation in bonding.

The cyclic ylid (1.38) appears to show normal carbanion character (p. 139) and hence presumably delocalization throughout the ring system does not occur. However, its benzene analogue (1.39), although sensitive to oxygen, is crystalline and can be prepared by treating the corresponding phosphonium salt (1.40) with aqueous sodium hydroxide (p. 136). As the related acyclic compound (1.41) is extremely sensitive to moisture and can only be prepared in solution, it is thought that (1.39) does show some delocalization.

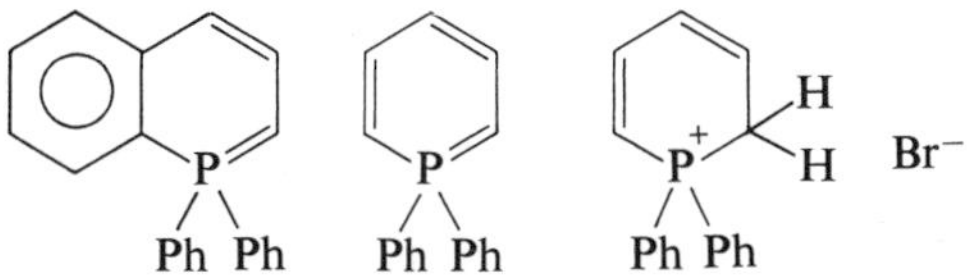

P Ph Ph (1.38) P Ph Ph (1.39) P^+ H H Ph Ph Br^- (1.40)

$$Ph_3P{=}CH—CH{=}CH—CH—CH_2$$

(1.41)

Two separate bonding schemes, both involving the use of d-orbitals, have been suggested to account for this extra stability. One uses the same d-orbital for bonding from phosphorus to both adjacent carbons (and so suggests complete delocalization through the phosphorus); the other uses a separate d-orbital for bonding to each adjacent carbon (and so suggests that the phosphorus atom will not allow conjugation through it). Insufficient evidence is available at present to choose between the two approaches.

Although it is impossible to say with certainty that d_{π} bonding takes place in phosphorus compounds, the postulate of bonding of this type conveniently accounts for a wide variety of observations and so its use seems justified until proof, or a superior alternative, becomes available.

1.4 Spectroscopy in phosphorus chemistry

The development and application of spectroscopic methods over the last twenty years have led to a wealth of new information in organic chemistry. Although attempts were made to use spectroscopy in phosphorus chemistry from the earliest opportunity, the information forthcoming was minimal because the special problems present in phosphorus chemistry were not taken into account. Spectroscopy in a fuller sense has only really been applied in the last few years and a great deal of further work will be necessary before the comprehensive correlations available for general organic chemistry are duplicated for the chemistry of phosphorus. However, an attempt has been made here to indicate the information available and some of the uses to which this information has been put. More specific information can be found throughout the text and a number of correlation charts are presented in the Appendix.

1.4.1 *Ultraviolet spectroscopy*

The indiscriminate use of spectroscopic correlations from general organic chemistry in the study of phosphorus compounds has been particularly widespread in ultraviolet spectroscopy. Attempts of this kind, with little or no consideration of the different symmetry of the orbitals used in bonding by phosphorus as compared with those used by carbon, have led to considerable confusion. Consequently, most early attempts at assignment of absorptions to specific transitions should be treated with suspicion.

The use of ultraviolet spectroscopy in the study of bonding involving d-orbitals has already been discussed (p. 36). It was pointed out that the symmetry of these orbitals is such that a node could exist at the phosphorus atom (i.e. the phosphorus atom might not pass on conjugation). This will be important, since in their simplest sense ultraviolet spectra provide a measure of conjugation.

Ultraviolet spectroscopy has been used, with varying success, to study the tautomerism of secondary phosphites, delocalization in the phosphazenes, and bonding and structure in a variety of phosphoryl compounds.

1.4.2 *Infrared spectroscopy*

The ranges of stretching frequency which have been developed for various functional groups in general organic chemistry are often still applicable when these groups are present in phosphorus compounds. The exceptions are where the functional group has undergone changes in bonding due to the presence of the phosphorus atom, for example, the carbonyl group absorption in the β-keto-ylid (1.42) is very weak and shifted to ~1600 cm^{-1}, presumably due to the resonance shown.

$$Ph_3P{=}CH{-}CO{-}CH_3 \longleftrightarrow Ph_3P^+{-}CH{=}\underset{\displaystyle O^-}{\underset{|}{C}}{-}CH_3$$

(1.42)

The stretching frequencies of most functional groups containing phosphorus have been determined (see Appendix) and are widely used in structural investigation of unknown compounds. The use of infrared spectroscopy in the study of bonding has already been discussed (pp. 36–7).

1.4.3 *Nuclear magnetic resonance*

Apart perhaps from mass spectrometry, nuclear magnetic resonance (n.m.r.) spectroscopy provides the greatest potential source of structural information in phosphorus chemistry. Considerable information can be obtained from the hydrogen-1 nuclear magnetic resonance spectra of phosphorus compounds and further information is available from phosphorus-31 nuclear magnetic resonance (^{31}P is the only natural isotope of phosphorus).

There are, however, some practical difficulties with ^{31}P n.m.r. measurements over and above those involving ^1H. The ^{31}P nucleus is more readily saturated than ^1H and so fairly rapid field sweeps are required to prevent promotion of all the nuclei to the higher energy level. Phosphorus-31 also shows a low nuclear sensitivity (only about 6 per cent of that shown by ^{19}F and ^1H); thus high concentrations and large samples are required. Because of this large volume it is more difficult to maintain a homogeneous magnetic field throughout the sample. Although the existence of phosphorus chemical shifts was recognized as early as 1949, difficulties of this nature have restricted the application of ^{31}P n.m.r. until fairly recently.

(a) *Chemical shift*. The ^{31}P nucleus is able to show resonance in an applied magnetic field in exactly the same way as the ^1H nucleus. However, the mean applied field at which this resonance takes place is very different for the two nuclei.

Whether resonance takes place or not for a particular hydrogen nucleus depends on the magnetic field experienced by that nucleus. This is not usually the same as the applied magnetic field because of the effects of the electrons in the vicinity of the nucleus, not only those associated with the nucleus being observed, but also those of adjacent nuclei. Thus, any change of electron density near the nucleus being studied will require a change in applied field if resonance is still to be observed. This is the origin of chemical shift.

The same is true of a phosphorus nucleus: whether it resonates or not depends on the field experienced by the nucleus. As phosphorus has fifteen times as many electrons as hydrogen it is reasonable to assume that in the *same* external magnetic field the two nuclei will experience very different fields. This is in fact the case and the difference in chemical shift between phosphorus and hydrogen is enormous compared with the difference between individual hydrogen nuclei (0·21 G covers the whole breadth of hydrogen chemical shifts at 14100 G, but phosphorus resonates at about 34800 G at the same frequency of 60 MHz).

The reference usually employed for ^{31}P spectra is 85 per cent aqueous orthophosphoric acid at 25 °C. Chemical shifts are then defined in the usual way as $\{(H_p - H_r)/H_r\} \times 10^6$ p.p.m., where H_p is the applied magnetic field for resonance of the sample and H_r that for the reference. The standard is usually external and this leads to some inaccuracy in determining relative sample peak position. Because of the large chemical shifts encountered in ^{31}P spectra (as large as 500 p.p.m.) this is less important than in ^{1}H spectra (where they rarely exceed 15 p.p.m.); an accuracy of ± 1 p.p.m. is usually sufficient. In those cases where the orthophosphoric acid is used as an internal standard, account must be taken of the change in its resonance position with pH of solution (a variation of about 6 p.p.m. takes place from H_3PO_4 to PO_4^{3-}).

Although trivalent phosphorus and pentavalent phosphorus have been shown to have very different chemical shifts, the close correlations of structure and chemical shift available in ^{1}H nuclear magnetic resonance are largely absent, at present, from ^{31}P spectra. A number of authors, notably Van Wazer and Schoolery, have attempted to develop correlations of this kind, but they are, in the main, unsatisfactory.

(b) *Spin–spin splittings.* Spin–spin splitting is an indirect coupling of two, or more, magnetic nuclei within the same molecule and is due to the modification of the magnetic field felt by one nucleus by the orientation of the nuclear moment of the other. The magnitude of this splitting is known as the coupling constant (J) between the two nuclei and is usually measured in hertz. In a complex spectrum it is sometimes difficult to decide whether a particular multiplicity of peaks is due to spin–spin splitting or chemical-shift differences. This problem is readily resolved by varying the applied magnetic field, because spin–spin splitting measured in hertz is independent of this, while chemical shift is not.

Spin decoupling, or double resonance, can also be used to simplify complex spectra. This involves the application of a frequency equal to the resonance frequency of one of the interacting nuclei while recording the resonance pattern of the other. The result of this is a complete removal of any coupling due to the first nucleus.

All these effects are observable, in theory, in interactions between any two atoms having a small quadrupole moment, but in practice observations of this type are confined to those which have no quadrupole moment, for example, ^{1}H, ^{19}F, ^{31}P. A further restriction on spin–spin splitting is that equivalent nuclei do not couple with each other, for example (1.43) shows only a single line in its n.m.r. spectrum.

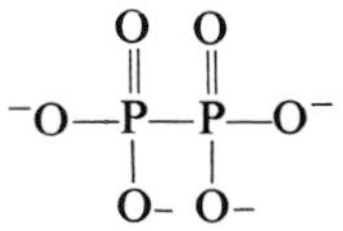

$$
\begin{array}{c}
\quad O \quad O \\
\quad \| \quad \| \\
{}^{-}O-P-P-O^{-} \\
\quad | \quad | \\
\quad O^{-} \; O^{-}
\end{array}
$$

(1.43)

A complete analysis of a spectrum involving the interaction of more than two non-equivalent nuclei is only possible without the use of the quantum-mechanical wave equation if the chemical shift difference between the nuclei is much greater than their spin–spin coupling. Fortunately, this is often the case even when the nuclei are of the same element and always the case when they are different elements. In a compound X_nY_m, Y will show $n+1$ resonance peaks and X will show $m+1$, for example, in phosphine, phosphorus shows a quartet of peaks while hydrogen shows a doublet. When a phosphorus nucleus interacts with a hydrogen the coupling can be transmitted over much greater distances than that between two hydrogen nuclei. For example, in the phosphorus ylid (1.44) the phosphorus atom shows a coupling with the methyl group attached to the γ-carbon atom ($J_{P-H} = 4$ Hz), whereas in hydrogen–hydrogen interactions,

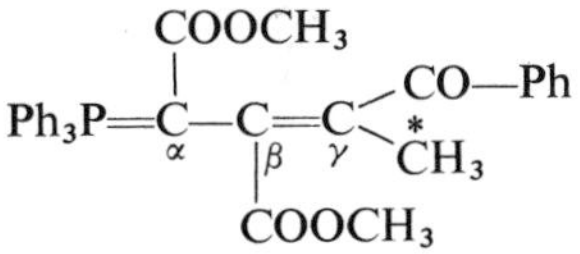

$$
Ph_3P{=}\underset{\alpha}{C}(COOCH_3)-\underset{\beta}{C}(COOCH_3){=}\underset{\gamma}{C}(CO{-}Ph)(\overset{*}{C}H_3)
$$

(1.44)

observable coupling does not extend beyond γ and usually not beyond β. Tables of coupling constants are given in the Appendix.

(c) *Applications of phosphorus-31 nuclear magnetic resonance.* The available correlations of chemical shift with structure have been used to distinguish trivalent and pentavalent phosphorus.

Ramirez has used chemical shifts to show that the adducts of trialkylphosphites and α-diketones have the structure (1.45) and not (1.46). The observed chemical shifts for these compounds are in the region of +50 p.p.m. and Ramirez suggests that betaine structure (1.46) is unlikely to provide shielding of this magnitude.

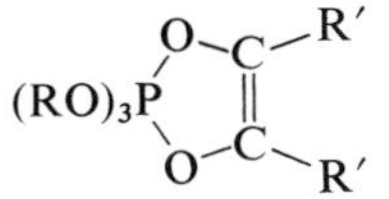

$$
(RO)_3P\langle{}^{O-C-R'}_{O-C-R'}\rangle \quad (C{=}C)
$$

(1.45)

$$
(RO)_3\overset{+}{P}-O-C(R'){=}C(R')-O^{-}
$$

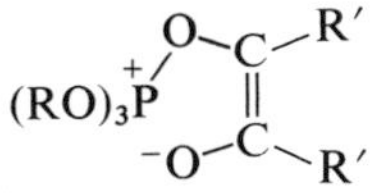

(1.46)

Phosphorus-31 chemical shifts have also been used to determine the equivalence, or otherwise, of phosphorus atoms in the same molecule. For example, in the phosphorus ylid $Ph_3P^{+}{-}CH{=}PPh_3Br^{-}$ the phosphorus atoms are equivalent and so the structure is really (1.47).

$$Ph_3P \overset{CH}{\overset{+}{\cdots}} PPh_3 \quad Br^-$$

(1.47)

Separate attempts to synthesize compounds with structures (1.48) and (1.49) were found to lead to the same product, the ^{31}P n.m.r. spectrum of which showed only one type of phosphorus atom, suggesting that the symmetrical structure (1.49) was correct. Phosphorus-31 resonance has also been used in this way to study the phosphazenes and condensed phosphorus acids.

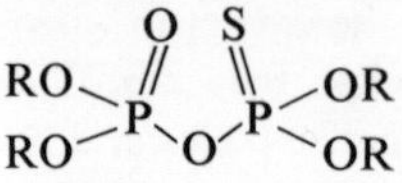

$$(RO)_2P(=O)-O-P(=S)(OR)_2 \qquad (RO)_2P(=O)-S-P(=O)(OR)_2$$

(1.48) 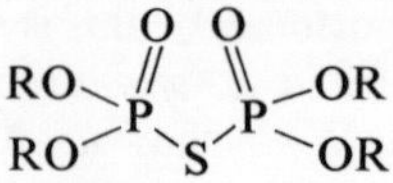 (1.49)

Spin–spin splitting due to phosphorus is widely used in a similar manner to spin–spin splitting due to hydrogen in structural determination. An important example of its use in this way was the final proof of the structure of secondary phosphites, which was shown to be largely (1.50) rather than (1.51) by the large coupling of phosphorus and hydrogen ($J_{P-H} \simeq 500$ Hz).

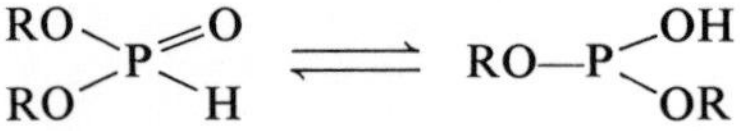

$$(RO)_2P(=O)H \rightleftharpoons RO-P(OH)(OR)$$

(1.50) (1.51)

Phosphorus-31 resonance is already an important tool in organic chemistry which is potentially of much greater application. Undoubtedly this will develop in the next few years.

1.4.4 *Mass spectrometry*

The enormous advance in the application of mass spectrometry to general organic-chemistry problems has not been paralleled in organophosphorus chemistry. Some studies have been made but, as with ^{31}P n.m.r. spectroscopy, the correlations available for general organic chemistry are often lacking for compounds containing phosphorus.

High-resolution mass spectrometry allows the determination of molecular weights with an accuracy of 1 in 10^6, and hence molecular formulae. This is widely used in phosphorus chemistry.

Low-resolution mass spectrometry involves a study of the fragments produced from the decomposition of a compound. The structure and abundance of these fragments, together with information available from similar decompositions, are extremely helpful in the determination of unknown structures. The number of published low-resolution mass spectra of phosphorus compounds is fairly small at present and this in turn restricts the information which can be obtained from any particular one. However, a combination of information from organic mass spectra and the known *chemical* stability of phosphorus-containing functional groups can often lead to very useful interpretations of phosphorus mass spectra.

1.4.5 *Other spectroscopic methods*

Microwave, electron spin resonance, Raman and a variety of other forms of spectroscopy have all been used in the study of phosphorus compounds, although in a fairly restricted way.

One similar physical technique which should be mentioned is X-ray diffraction. This has been used in the determination of some particularly awkward and important organophosphorus structures.

Problems

1.1 Outline the main differences between the organic chemistries of phosphorus and nitrogen, and suggest an explanation for these differences.
[See Chapter 1, pp. 15–27 and Hudson (1965), pp. 8–39.]

1.2 The bond order (the number of two-electron bonds) of the phosphoryl (PO) group can be calculated from force-constant data. Explain the relative magnitude of the bond orders listed below.

	Bond order
PO_4^{3-}	1·50
$F_2PO_2^-$	1·84
$P(O)Cl_3$	1·95
$P(O)FCl_2$	2·05
$P(O)F_3$	2·22

1.3 5,10-Dihydro-5,10-diethylphosphenanthrene (1.52) can be isolated in two isomeric forms. Explain.

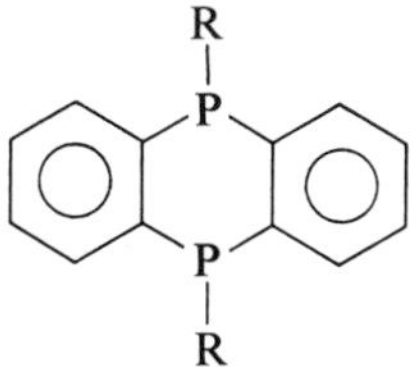

R
P
P
R

(1.52)

[Kennard, Mann, Watson, Fawcett and Kerr (1968).]

1.4 Inversion barriers for phosphines are normally in the region of 126 kJ mol^{-1} (30 kcal mol^{-1}) and so not readily observable by n.m.r. because the temperature required for inversion is too high. However, the barrier to inversion at phosphorus in the cyclic phosphine (1.53) is readily measured by n.m.r. and has been shown to be only 67 kJ mol^{-1} (16 kcal mol^{-1}). Why should this barrier be so low?

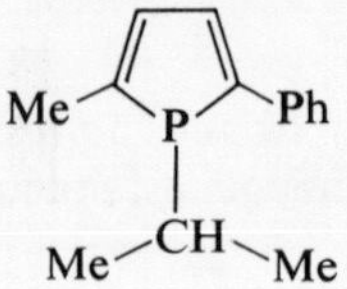

Me P Ph
Me CH Me
(1.53)

[Egan, Tang, Zon and Mislow (1970).]

1.5 N.M.R. studies suggest that the barrier to inversion at phosphorus in diphosphines (e.g. 1.54) is considerably less than the barrier in monophosphines (e.g. 1.55). Why should this be the case?

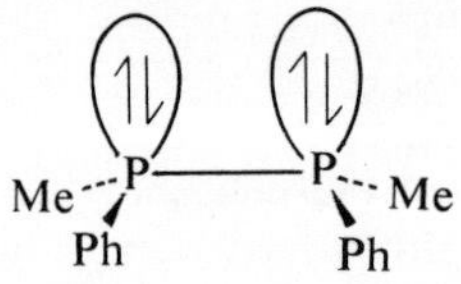

Me P P Me
Ph Ph
(1.54)

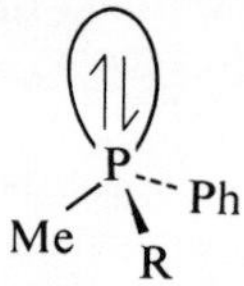

P Ph
Me R
(1.55)

[Lambert, Jackson and Mueller (1968).]

1.6 The characteristic infrared stretching frequency for P=O usually lies in the region 1240–1260 cm^{-1} but in compounds (1.56) and (1.57) this is shifted to near 1200 cm^{-1}. Explain.

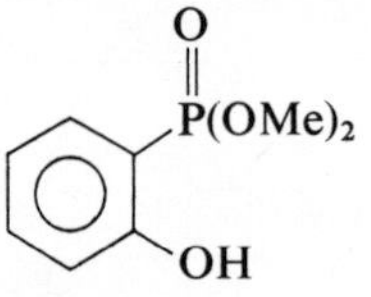

O
P(OMe)$_2$
OH
(1.56)

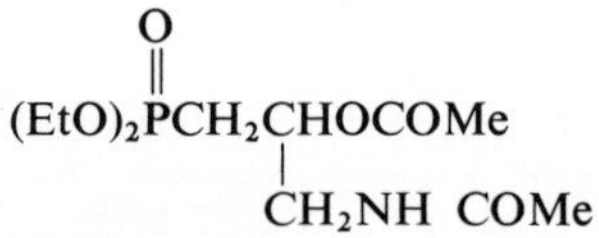

O
(EtO)$_2$PCH$_2$CHOCOMe
CH$_2$NH COMe
(1.57)

[Obrycki and Griffin (1968).]

1.7 The compound $Et_2N—PF_4$ is found to have temperature-dependent ^{19}F and ^{31}P n.m.r. spectra. At room temperature the ^{19}F n.m.r. shows a single doublet due to spin coupling with phosphorus. At low temperatures the ^{19}F n.m.r. spectrum shows two doublets, each with half the intensity of the doublet observed at room temperature. Each component of these doublets shows extra triplet fine structure.

From this information suggest a possible structure for Et_2NPF_4 and in terms of this structure explain the observations made.
[Schmutzler (1965), p. 506.]

Chapter 2
Trivalent-Phosphorus Compounds

Compounds in this category have the general formula R_3P, where R can be hydrogen, alkyl, aryl, alkoxy, halogen, $—NR_2$, —S—R, $—SiR_3$, monovalent metal. Trivalent-phosphorus compounds that contain one or more hydroxyl groups will be discussed in Chapter 3 as they exist largely in the phosphoryl form (2.1) and their properties reflect this.

$$R_2P—OH \rightleftharpoons \underset{(2.1)}{R_2P(=O)—H}$$

The range of reactivity in trivalent-phosphorus compounds is enormous. For example, although they are generally susceptible to oxidation, compounds containing P—H bonds are dangerously so (ether solutions of $PhPH_2$ are spontaneously inflammable), while aminophosphines and halophosphines can often be exposed to dry air without appreciable oxidation.

Bond angles in trivalent-phosphorus compounds are generally smaller than those in the corresponding nitrogen compounds. This is due to the difference in hybridization between phosphorus, which has three p-orbitals with a small amount of s-character and one s-orbital with a small amount of p-character containing the lone pair, and nitrogen, which is sp^3 hybridized. We shall see how these differences affect the behaviour of the corresponding compounds (pp. 49 ff.).

Many of the reactions of trivalent-phosphorus compounds involve conversion into phosphorus(V). When the phosphorus acts as a nucleophile, a four-coordinated phosphorus cation (e.g. 2.2) is initially formed and in many cases is the isolated product (see phosphonium salts, p. 127). This is not true in trivalent-phosphorus compounds containing a P—O bond (e.g. p. 56), whose chemistry tends to be much more complex.

$$R^1I + R_3P: \longrightarrow \underset{(2.2)}{R_3P^+—R^1 \quad I^-}$$

The wide range of reactivity of trivalent-phosphorus compounds makes classification difficult. The mechanistic categorization used in this chapter has been adopted with a view to promoting a real understanding of the extent of, and reasons for, this range.

Table 11 Nomenclature of trivalent-phosphorus compounds

R_3P:	trialkylphosphine
Phosphinous acid derivatives	
$R_2P—OH$	dialkylphosphinous acid
$R_2P—OR'$	alkyl dialkylphosphinite
$R_2P—Cl$	dialkylphosphinous chloride
$R_2P—NR'_2$	*N*,*N*-dialkyl dialkylphosphinous amide
$R_2P—SR'$	*S*-alkyl dialkylphosphinothioite
Phosphonous acid derivatives	
$RP(OH)_2$	alkylphosphonous acid
$RP(OR')_2$	dialkyl alkylphosphonite
$RPCl_2$	alkylphosphonous chloride
$RP(NR'_2)_2$	*N*,*N*,*N*,*N*-tetraalkyl alkylphosphonous diamide
$RP(SR')_2$	*S*,*S*-dialkyl alkylphosphonodithioite
Phosphorous acid derivatives	
$(HO)_3P$:	phosphorous acid
$(RO)_3P$	trialkyl phosphite
Cl_3P	phosphorus trichloride
$(R_2N)_3P$	hexaalkylphosphorous triamide
$(RS)_3P$	trialkyl phosphorotrithioite

A summary of nomenclature for trivalent-phosphorus compounds is given in Table 11 and is aimed at reducing any pain the reader might feel in connection with this topic. However, formulae and descriptive names used extensively throughout the text should make the use of this table minimal.

2.1 Nucleophilic reactivity

Trivalent-phosphorus compounds contain a lone pair of electrons and so would be expected to show nucleophilic character in their reactions. They show this behaviour in reactions at both electron-deficient centres (e.g. $>C=O$, $\geq P=O$) and electron-rich centres like oxygen and the halogens.

Although phosphines are generally weaker bases than amines they are invariably stronger nucleophiles. Much has been written about nucleophilicity (see the Bibliography) and so let it suffice for us to say that the differences between nitrogen and phosphorus are mainly due to the higher polarizability of phosphorus making its electron pair more readily available for reaction. This same higher polarizability allows phosphines, but not amines, to attack centres of high electron density by reducing electronic repulsion.

Trivalent-phosphorus compounds can also act as bases, for example the reaction of triphenylphosphine with tertiary butyl chloride in refluxing formic acid gives 2-methyl-1-propene (2.3) in good yield, presumably by the mechanism

Table 12 Reactions involving trivalent phosphorus acting as a base

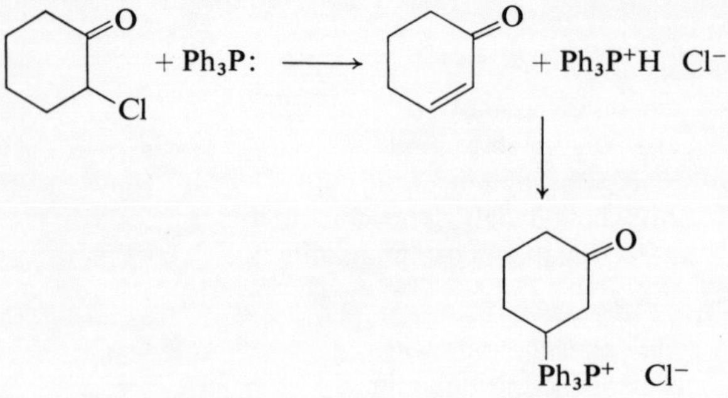

O, Cl $+ Ph_3P: \longrightarrow$ O $+ Ph_3P^+H\ Cl^-$

$\downarrow$

O, $Ph_3P^+\ Cl^-$

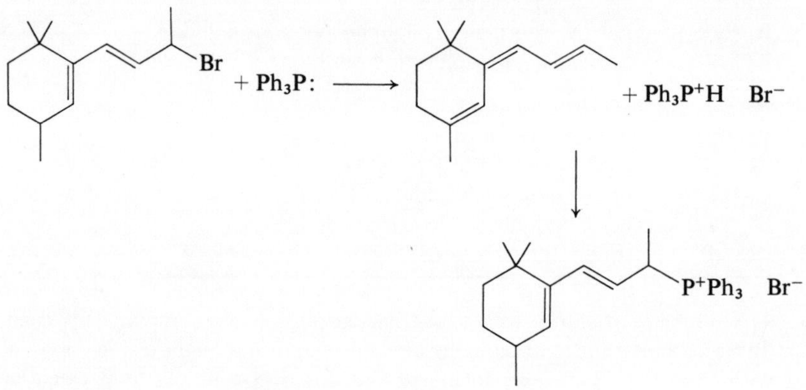

Br $+ Ph_3P: \longrightarrow$ $+ Ph_3P^+H\ Br^-$

$\downarrow$

$P^+Ph_3\ Br^-$

$PH_3 + HI \longrightarrow PH_4^+\ I^-$

(This is the only example of a crystalline PH_4^+ salt.)

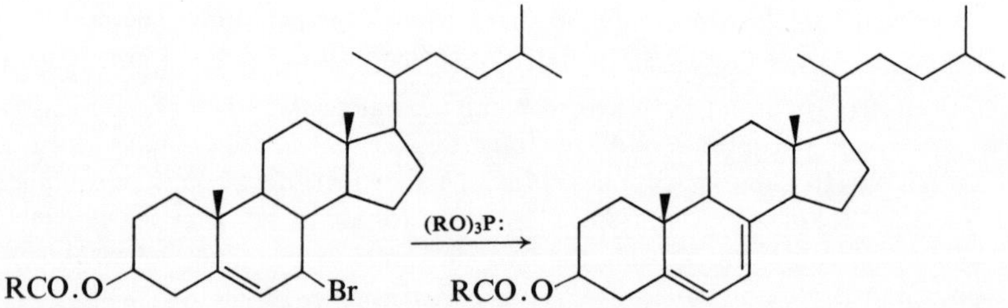

RCO.O, Br $\xrightarrow{(RO)_3P:}$ RCO.O

shown. However this only occurs in special cases (see Table 12) like tertiary halides, and nucleophilic displacement usually predominates.

$$(CH_3)_3CCl \rightleftharpoons (CH_3)_3C^+ + Cl^-$$

$$Ph_3P\!: + H{-}CH_2{-}\overset{\overset{CH_2}{|}}{\underset{\underset{CH_2}{|}}{C^+}} \longrightarrow Ph_3P^+H + CH_2{=}C\begin{matrix}CH_3\\CH_3\end{matrix}$$

(2.3)

Let us now consider various reactions in which trivalent-phosphorus compounds act as nucleophiles.

2.1.1 *Substitutions*

The most common examples of substitution reactions in organic chemistry are where the displacement takes place by attack at carbon (e.g. the hydrolysis of alkyl halides 2.4), but with phosphorus nucleophiles the substitution can take place at a wide variety of different atoms (e.g. at halogen).

$$HO^- \; \rangle C{-}X \longrightarrow \; \rangle C{-}OH + X^-$$

(2.4)

$$R_3P\!: \; Br{-}Br \longrightarrow R_3P^+Br \quad Br^-$$

What controls the centre of attack in circumstances where the nucleophile has a choice? Bond energies are often very important, for example phosphorus–halogen bonds (see p. 18) are generally stronger than oxygen–halogen or nitrogen–halogen bonds; consequently a much stronger bond is formed by a phosphorus nucleophile attacking a halogen centre than by an oxygen or nitrogen nucleophile. We have already seen that polarizability (the extent to which the non-bonded electrons of the nucleophile can be deformed) is very important in this type of reaction and trivalent phosphorus atoms are readily polarizable. Substituents will affect this polarizability to some extent as well as influencing the site of reaction by steric effects. Finally the centre of attack may be controlled by the nature of the group being displaced; for example, in ethyl iodide, nucleophilic attack at carbon will displace an iodide anion, a far better leaving group than the ethyl anion which an attack on iodine would displace.

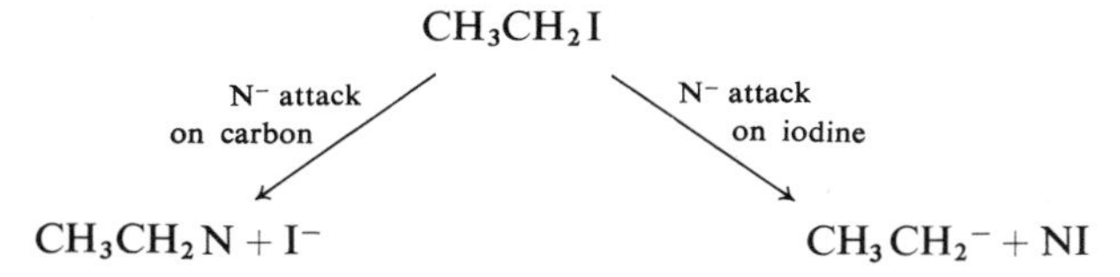

We must also consider the stereochemistry of these substitutions. The geometry at the centre of displacement will depend on the mechanism of substitution (i.e. largely inversion for S_N2 mechanisms and racemization for S_N1 mechanisms). The geometry about the phosphorus atom appears to be unaffected and, if an optically active phosphine is used, its configuration is retained. This seems quite reasonable since phosphines have stable configurations at temperatures below 100 °C (see p. 26) and the mechanism of substitution suggests no reason for inversion or racemization.

$$R^1R^2R^3P: + \geq C{-}X \rightarrow \left[R^1R^2R^3P{-}{-}{-}C{-}{-}{-}X \right]^{\ddagger} \rightarrow R^1R^2R^3P^+{-}C\leq \quad X^-$$

Table 13 Nucleophilic substitution by trivalent phosphorus: displacement at carbon

$$Ph_3P + CH_3Br \xrightarrow[5\,°C]{benzene} Ph_3P^+{-}CH_3 \quad Br^-$$
(m.p. 227–9 °C)

$$Ph_3P + PhCH_2Br \xrightarrow[reflux\ 20\ min]{benzene} Ph_3P^+{-}CH_2{-}Ph \quad Br^-$$

1-ethyl-4-hydroxy-4-(2-bromoethyl)phosphorinane (HO, CH_2CH_2Br; P–CH_2CH_3) $\xrightarrow[80\,°C]{benzene}$ bicyclic phosphonium salt (OH; P^+–CH_2CH_3) Br^-

o-(Et_2P)C_6H_4–CH_2–CH_2–CH_2–O–CH_3 $\xrightarrow[reflux]{HBr(aq.)}$ cyclic phosphonium ($\overset{+}{P}Et_2$) Br^-

$$Bu_3P + CH_3CH_2O{-}C^+(OEt)_2\ BF_4^- \longrightarrow Bu_3P^+{-}CH_2CH_3\ BF_4^- + (EtO)_2C{=}O$$

$$Ph_2P^-Na^+ + CH_3(CH_2)_3Br \xrightarrow[THF]{RT} Ph_2P(CH_2)_3CH_3 + NaBr$$
(61 %)

$$(C_6H_{11})_2P^-Li^+ + CH_2{-}CH_2\ (\text{epoxide, O}) \xrightarrow[(ii)\ H_2O]{(i)\ dioxan\ RT} (C_6H_{11})_2P{-}CH_2CH_2OH$$
(37%; m.p. 110–12 °C)

Table 13 – continued

$$EtPLi_2 + X(CH_2)_nX \longrightarrow EtP\widehat{(CH_2)_n} \quad (X = \text{halogen}; n = 4 \text{ or } 5)$$

$$(n = 4; 22\%)$$

$$\begin{array}{c} RPLi-CH_2 \\ | \\ RPLi-CH_2 \end{array} + \begin{array}{c} Cl-CH_2 \\ | \\ Cl-CH_2 \end{array} \longrightarrow \begin{array}{c} RP(CH_2CH_2)_2PR \end{array} + 2LiCl$$

$$2H_2P^-Na^+ + Cl—CH_2—CH_2—Cl \xrightarrow{\text{liquid } NH_3} \underset{\underset{H}{P}}{CH_2—CH_2} + PH_3 + CH_2{=}CH_2$$

$$(EtO)_3P + CH_3I \xrightarrow[\text{reflux 2h}]{\text{benzene}} (EtO)_2\overset{O}{\overset{\|}{P}}CH_3 + EtI$$

$$MeOPPh_2 + PhCH_2Br \xrightarrow{120\,°C} Ph_2\overset{O}{\overset{\|}{P}}CH_2Ph + CH_3Br$$

(a) *Substitution at carbon.* Probably the best-known substitution reaction of phosphines is the formation of phosphonium salts (2.5) with alkyl halides (see Table 13).

$$R_3P\!: + R'—CH_2X \xrightarrow{\text{benzene}} \underset{(2.5)}{R_3P^+—CH_2—R' \quad X^-}$$

The reaction can be carried out in a variety of solvents, although benzene is most commonly used. It takes place readily with primary, and many secondary, halides with an order of reactivity of I > Br > Cl. The mechanism appears to be largely S_N2 in most cases as reactions with optically active alkyl halides lead to inversion of configuration at the asymmetric carbon atom.

$$R_3P\!: + \underset{R}{\overset{H\ \ CH_3}{C}}—X \longrightarrow \left[R_3P^{\delta+} \cdots \underset{R}{\overset{H\ \ CH_3}{C}} \cdots X^{\delta-} \right]^{\ddagger} \longrightarrow R_3P^+—\underset{R}{\overset{CH_3\ \ H}{C}} \quad X^-$$

Tertiary halides will not usually undergo substitution reactions with tertiary phosphines, although the salts of tertiary halides can sometimes be prepared from the corresponding alcohol under acid conditions. This suggests an S_N1 mechanism for these reactions.

$$R_3C{-}OH + H^+ \longrightarrow R_3C{-}O^+H_2 \rightleftharpoons R_3C^+ + H_2O$$

$$\downarrow R'_3P:$$

$$R'_3P^+{-}CR_3$$

When the phosphorus nucleophile contains a phosphorus–oxygen bond different products are obtained. Alkyl halides undergo an initial substitution reaction with trialkyl phosphites $(RO)_3P$ to give the phosphite equivalent of a phosphonium salt (2.6). However the salt (2.6) is unstable in the presence of

$$(CH_3O)_3P: + CH_3CH_2I \longrightarrow (CH_3O)_3P^+{-}CH_2CH_3 \quad I^- \quad (2.6)$$

$$(CH_3O)_2P^+(OCH_3){-}CH_2CH_3 \;\; I^- \longrightarrow (CH_3O)_2P(=O){-}CH_2{-}CH_3 + CH_3I \quad (2.7)$$

nucleophiles and is rapidly attacked by iodide ion at one of the methoxy carbon atoms to give a phosphonate (2.7). The reaction of trialkyl phosphites with alkyl halides to give phosphonates is known as the *Michaelis–Arbusov* reaction. This reaction is important since phosphonates are very useful in the synthesis of olefins (p. 118).

The corresponding phosphinites $ROPR_2$ and phosphonites $(RO)_2PR$ undergo identical reactions to give phosphine oxides (2.8) and phosphinates (2.9) respectively.

$$R'OPR_2 + CH_3CH_2I \longrightarrow R'O{-}P^+R_2{-}CH_2CH_3 \quad I^-$$

$$\downarrow$$

$$R'I + R_2P(=O){-}CH_2CH_3 \quad (2.8)$$

$$(R'O)_2PR + PhCH_2Br \longrightarrow R'O\text{—}\overset{\overset{R'}{|}}{\underset{\underset{R}{|}}{\overset{O}{\overset{|}{P^+}}}}\text{—}CH_2Ph\ \ Br^-$$

$$\downarrow$$

$$R'O\text{—}\overset{\overset{O}{\|}}{\underset{\underset{R}{|}}{P}}\text{—}CH_2Ph + R'Br$$

(2.9)

There are many reactions of the Michaelis–Arbusov-type in organophosphorus chemistry (for some examples see Table 13), the driving force in all cases being the formation of the extremely strong P=O bond. Trivalent-phosphorus esters will themselves rearrange in a reaction which is presumably related to the Michaelis–Arbusov. The product (2.10) again contains a P=O group.

$$(RO)_2P\text{—}OR \xrightarrow{\text{heat}} (RO)_2\overset{\overset{O}{\|}}{P}\text{—}R$$

(2.10)

This reaction only takes place easily when alkyl phosphites are used. Triphenyl phosphite reacts with alkyl halides to give a salt which only decomposes on heating to more than 200 °C.

$$(PhO)_3P{:} + RI \longrightarrow (PhO)_3P^+\text{—}R\quad I^-$$

The need for the higher temperature is explained in terms of the second stage in the reaction, which involves the attack of a nucleophile on an aromatic nucleus, a very slow process unless the ring is especially activated.

$$C_6H_5\text{—}O\text{—}\underset{(OPh)_2}{P^+}\text{—}R\ \ I^- \longrightarrow C_6H_5\text{—}I + (PhO)_2\overset{\overset{O}{\|}}{P}R$$

The Michaelis–Arbusov reaction is less common with phosphorotrithioites $(RS)_3P$ and does not take place at all with alkylphosphorous triamides $(R_2N)_3P$, where the expected product would be the phosphine imide (2.11). This may be

$$(R_2N)_2\underset{\underset{\underset{X^-\ R}{|}}{NR}}{\overset{}{\underset{|}{P^+}}}\text{—}R' \longrightarrow (R_2N)_2\underset{\underset{NR}{\|}}{P}\text{—}R'$$

(2.11)

due to the difference in bond energies between P=O (627 kcal mol^{-1}, 150 kcal mol^{-1}) and P=N (459 kJ mol^{-1}, 110 kcal mol^{-1}).

Epoxide rings are opened by both phosphines and phosphites to give the intermediate betaine (2.12). When derived from phosphines, betaines of this

$$R_3P: + \underset{\underset{R}{|}}{CH}\!-\!CH_2\!-\!O \longrightarrow R_3P^{+}\!-\!CH_2\!-\!\overset{O^-}{\overset{|}{C}H}\!-\!R \quad (2.12)$$

type decompose to give olefins and phosphine oxides (see p. 142; the Wittig reaction). However, betaines formed from phosphites undergo a type of intramolecular Michaelis–Arbusov reaction to give phosphonates (2.13).

$$(R'O)_3P: + \underset{\underset{R}{|}}{CH}\!-\!CH_2\!-\!O \longrightarrow (R'O)_2\overset{O-R'}{P^{+}}\!-\!CH_2\!-\!\underset{\underset{R}{|}}{CH}\!-\!O^-$$

$$\downarrow$$

$$(R'O)_2\overset{O}{\overset{\|}{P}}\!-\!CH_2\!-\!\underset{\underset{R}{|}}{CH}\!-\!OR' \quad (2.13)$$

(b) *Substitution at halogen.* Nucleophilic attack of phosphorus at halogen centres is far more common than with carbon nucleophiles. Some of the reasons for this have already been discussed (p. 51).

(i) Halogen molecules. Both phosphines and phosphites react readily with halogens, the former by nucleophilic substitution to give an addition compound which can be represented by an ionic (2.14a) or a covalent (2.14b) formulation.

$$Ph_3P: + Br\!-\!Br \longrightarrow \begin{cases} Ph_3P^{+}Br \quad Br^{-} & (2.14a) \\ Ph_3PBr_2 & (2.14b) \end{cases}$$

The evidence available suggests that, depending on the substituents, either state can exist. Support for participation from (2.14b) is provided by the ready racemization of optically active phosphines with traces of bromine, which can only take place through the covalent form (2.14b) (see p. 31).

The reactions of phosphites, phosphinites, etc. depend on the substituents. Triaryl phosphites react in a similar way to phosphines to give addition products (2.15a or b) initially, although this is followed by some disproportionation.

$$(ArO)_3P + Br_2 \longrightarrow \left\{ \begin{array}{l} (ArO)_3PBr_2 \\ (2.15a) \\ \\ (ArO)_3P^+Br \quad Br^- \\ (2.15b) \end{array} \right\} \xrightarrow{(ArO)_3P:} (ArO)_4PBr + (ArO)_2PBr$$

Trialkyl phosphites (2.16), on the other hand, undergo a reaction analogous to the Michaelis–Arbusov reaction (see p. 54).

$$(RO)_3P: + Br—Br \longrightarrow (RO)_2P^+—Br \longrightarrow (RO)_2P(=O)Br + RBr$$

(2.16) $\quad$ O—R $\quad Br^-$

(ii) Carbon–halogen bonds. In these compounds the phosphorus nucleophile may attack either at a carbon atom with displacement of a halide ion, or at a halogen atom with displacement of a carbanion. We have already looked at the first of these possibilities (see p. 53).

Before extensive studies of phosphorus chemistry took place, most known examples of substitution reactions of this type involved attack at a carbon atom. When phosphorus nucleophiles were, in many cases, found to attack halogen, these reactions were called *abnormal* and this name has stuck. Of course attack on halogen is no more abnormal than attack on any other centre, though it is generally less frequent with nucleophiles from the first row of the periodic table (see p. 51).

Numerous examples of nucleophilic attack on the halogen of carbon–halogen bonds are now known. The ease of this attack follows the order I > Br > Cl for reasons of polarizability, and compounds which undergo this reaction are said to have *activated* or *positive* halogens. Activated halogens arise when the carbanion which is formed by displacement at a halogen atom is stabilized, either by resonance, as in α-halo-ketones, -esters, -nitriles, etc., or by inductive effects, as in polyhalogen compounds like $BrCF_3$.

The products of these reactions depend on the reaction conditions, and it is easy to be misled by their apparent derivation from attack at some centre other than the halogen atom. However, the first stage in all these reactions is the same: displacement occurs at a halogen atom to give an ion pair (2.18). This ion pair can then decompose in a number of different ways.

The carbanion can displace bromide ion by attack at a phosphorus atom to give the phosphonium salt (2.19). This salt is identical to that which would be formed in one step by the attack of phosphorus at a carbon atom. In early work, this similarity led to considerable confusion, which was only resolved when it was found that the ion pair (2.18) formed initially in the *abnormal* reactions

$$R_3P: + Br\text{—}CH(R')\text{—}CO\text{—}CH_3 \longrightarrow R_3P^+Br \quad R'\text{—}C^-H\text{—}CO\text{—}CH_3$$

(2.17) (2.18)

$$R_3P: + CH(Br)(R')\text{—}CO\text{—}CH_3 \longrightarrow R_3P^+\text{—}CH(R')\text{—}CO\text{—}CH_3Br^-$$

(2.17) (2.19)

reacted rapidly with hydroxylic compounds, for example methanol, to give debrominated ketone (2.20) and phosphine oxide (2.21).

$$H\text{—}\ddot{O}\text{—}CH_3 \quad R_3PBr^+ \quad H(OCH_3)\ C^-H(R)\text{—}CO\text{—}CH_3 \longrightarrow R_3P^+\text{—}O\text{—}CH_3\ Br^- + CH_2(R)\text{—}CO\text{—}CH_3 \longrightarrow R_3P{=}O + CH_3Br$$

(2.18) (2.20) (2.21)

Thus the addition of methanol, or water, to a reaction of this type provides a diagnostic test for the mechanistic pathway. Formation of the phosphonium salt (2.19) under these conditions confirms that the attack was at a carbon atom, while the production of major amounts of phosphine oxide and debrominated ketone suggests the attack occurred at a halogen atom, with the formation of an ion pair (2.18). Further confusion is caused by the many examples of reactions in which both mechanisms take place simultaneously, the importance of each mechanism depending on the temperature, solvent, and so on. The addition of the hydroxylic solvent may itself change the reaction pathway, so it is not certain that the reaction being diagnosed is the same as the original reaction in a non-hydroxylic solvent.

The anion in the ion pair is often *ambident*. This means that it has its negative charge spread by resonance and so any electrophile will have a choice in its position of attack. The anion in (2.18) is a resonance hybrid with negative charge spread over both oxygen and carbon (2.22).

$$R\text{—}^-CH\text{—}C({=}O)\text{—}CH_3 \longleftrightarrow R\text{—}CH{=}C(O^-)\text{—}CH_3$$

(2.22)

We have already discussed the products derived from attack by the positive phosphorus centre at the carbon of this resonance hybrid. When the attack is at oxygen a phosphonium salt (2.23) is still formed but it is an *enol-phosphonium* salt and contains a phosphorus–oxygen bond.

$$R_3P^+\!-\!Br \quad O^- \quad R\!-\!CH\!=\!C(O^-)\!-\!CH_3 \longrightarrow R_3P^+ \; Br^- \quad R\!-\!CH\!=\!C(O\!-\!P^+R_3)\!-\!CH_3 \; (2.23)$$

With α-halo-esters, -nitriles and -amides both mechanisms (i.e. the initial attack of phosphine at either a halogen or a carbon atom) probably still operate, but, unlike the reactions with α-haloketones, only those products analogous to (2.19) with phosphorus–carbon bonds are obtained. Enol-phosphonium salts analogous to (2.23) are not found in these cases, presumably because the ester, nitrile and amide anions do not show sufficient charge density at oxygen or nitrogen.

The above reactions take place with trivalent-phosphorus compounds containing phosphorus–carbon bonds (phosphines) and phosphorus–nitrogen bonds (aminophosphines). Phosphorus nucleophiles containing phosphorus–oxygen bonds (phosphites, phosphinites, phosphonites) can also attack activated halogen compounds (e.g. 2.24) at either halogen or carbon atoms. Both routes lead to the same product, a phosphonate (2.26), via an intermediate salt (2.25) and an Arbusov reaction.

$$(RO)_3P\!: + Br\!-\!CH(R')\!-\!C(=O)\!-\!O\!-\!Et \; (2.24) \longrightarrow (RO)_3P^+\!-\!Br \quad {}^-CH(R')\!-\!C(=O)\!-\!OEt$$

$$(RO)_3P\!: + CH(CO\!-\!OEt)(R')\!-\!Br \; (2.24) \longrightarrow (RO)_2P^+(O\!-\!R)\!-\!CH(R')\!-\!C(=O)\!-\!O\!-\!Et \quad Br^- \; (2.25)$$

$$\xrightarrow{\text{Arbusov}} (RO)_2P(=O)\!-\!CH(R')\!-\!C(=O)\!-\!O\!-\!Et \; (2.26) \; + RBr$$

Alkyl phosphites can also react with activated halogen compounds to give enol-phosphates (2.27) (the Perkov reaction). It was originally thought that

$$(RO)_3P: + X{-}\underset{\displaystyle R'}{\underset{|}{C}}H{-}\overset{\displaystyle O}{\overset{\|}{C}}{-}R' \longrightarrow (RO)_2\overset{\displaystyle O}{\overset{\|}{P}}{-}O{-}\underset{\displaystyle R'}{\underset{|}{C}}{=}CHR' + RX$$

(2.27)

(2.27) was formed by an attack on halogen in a way analogous to the formation of an enol-phosphonium salt (2.24), but recently it has been shown that the mechanism is completely different and involves attack on the carbonyl group. In view of this, the Perkov reaction will be discussed in more detail below (p. 68).

Both phosphines and phosphites react with polyhaloalkanes: the phosphine reaction almost certainly involves an attack on a halogen atom, while the phosphite reaction appears to be free radical in some instances. Triphenylphosphine reacts with carbon tetrachloride (or tetrabromide) to give a phosphorus ylid (2.28) (see p. 138), probably by the following mechanism.

$$Ph_3P: \; X{-}CX_3 \longrightarrow Ph_3P^+{-}X \quad C^-X_3$$

$$\downarrow$$

$$Ph_3P^+{-}CX_2{-}X \quad :PPh_3$$

$$\downarrow$$

$$Ph_3P^+{-}C^-X_2 + Ph_3PX_2$$

(2.28)

(iii) Other halogen bonds. Trivalent-phosphorus nucleophiles will attack the halogen atom in a number of other types of molecule.

$$(CH_3)_3C{-}O{-}Cl + :PPh_3 \longrightarrow (CH_3)_3C{-}O^- + Ph_3P^+Cl$$

$$\downarrow$$

$$Cl^- \quad (CH_3)_3C{-}O{-}P^+Ph_3$$

$$\downarrow$$

$$(CH_3)_3CCl + Ph_3P{=}O$$

(2.29)

Tertiary-butyl hypochlorite reacts with triphenylphosphine by initial attack on the halogen atom to give triphenylphosphine oxide (2.29) and tertiary-butyl chloride as the final products. The decomposition of the enol-phosphonium salt may well be by an S_N1 mechanism although it is represented here by S_N2.

Nitrogen–halogen and sulphur–halogen bonds are also attacked at the halogen atom under some circumstances. Examples of all these reactions can be found in Table 14.

Table 14 Nucleophilic substitution by trivalent phosphorus: displacement at halogen

$$Ph_3P^+{-}CH_2{-}Br\ Br^- + :PPh_3 \longrightarrow Ph_3P^+{-}C^-H_2 + Ph_3P^+Br\ Br^-$$

2-Bromoindane-1,3-dione (O, O, —Br) + :PPh₃ ⟶ enol-phosphonium salt (O—P⁺Ph₃, O) Br⁻

$$R_2P{-}PR_2 + Br{-}Br \longrightarrow R_2P{-}P^+(Br)R_2\ Br^- \longrightarrow 2R_2PBr$$

Hexachlorocyclopentadiene (Cl, Cl, Cl, Cl, Cl, Cl) + $(EtO)_3P:$ ⟶ 5-ethyl-pentachlorocyclopentadiene (Cl, Cl, Cl, Cl, Et, Cl) + $(EtO)_2P(=O)Cl$

$$(EtO)_3P: + Cl_2 \longrightarrow (EtO)_2P(=O)Cl + EtCl$$

R_3P + N-bromosuccinimide (O, N—Br, O) ⟶ R_3P^+Br succinimide anion (O, N⁻, O)

$$(RO)_3P: + Cl{-}S(=O)_2{-}Cl \longrightarrow (RO)_2P(=O)Cl + RCl + SO_2$$

(c) *Substitution at other centres.* Phosphorus nucleophiles can carry out substitutions at centres other than carbon and halogen, the most common of these being oxygen and sulphur. However, in many cases the mechanisms are rather uncertain and free radicals may be involved.

Tertiary phosphines reduce ozonides to give aldehydes, or ketones, and tertiary phosphite oxides. The use of oxygen-18 labelled ozonides (2.30) has demonstrated that the initial attack is at peroxide oxygen.

Ph_3P: ... Ph_3P^+ ... $Ph_3P{=}O$ + $>C{=}O^{18}$ + $>C{=}O$

(2.30)

Peroxides react readily with trivalent-phosphorus compounds of all types to give phosphorus compounds containing P═O bonds and reduction products of the peroxides. Although some of these reactions are undoubtedly free radical, there are cases where ionic mechanisms operate. Triphenylphosphine reacts with *n*-dialkyl peroxides by nucleophilic attack on an oxygen atom to give an alkoxyphosphonium salt intermediate (2.31) which decomposes to give phosphine oxide and a dialkyl ether by a route analogous to the Arbusov reaction. A similar reaction takes place with hydroperoxides (R—O—O—H) to give alcohols.

$$R{-}O{-}O{-}R + Bu_3P: \longrightarrow \left[Bu_3P^+{-}O{-}R \quad {}^-OR \right] \longrightarrow Bu_3P{=}O + ROR$$

(2.31)

Trivalent-phosphorus compounds can also react with disulphides and again the mechanism could be free radical or ionic. However the S—S bond is some 84 kJ mol^{-1} (20 kcal mol^{-1}) stronger than the O—O bond in peroxides, so ionic mechanisms are probably more common with sulphur compounds. Trialkyl phosphites can react with disulphides by either a homolytic or an ionic mechanism but, in the presence of an inhibitor, for example hydroquinone, the route is ionic. The products are the thiophosphate (2.32) and a mixed thioether. That the second step of the reaction involves an Arbusov-type reaction with the formation

$$BuS{-}SBu + (EtO)_3P \xrightarrow{\text{hydroquinone}} (EtO)_3P^+{-}S{-}Bu \quad BuS^-$$

$$\downarrow$$

$$(EtO)_2P(=O){-}S{-}Bu + BuSEt$$

(2.32)

of a P=O rather than P=S bond probably reflects the different strengths of these bonds, since both pathways are available (see p. 34).

All trivalent-phosphorus compounds react with atmospheric oxygen or elemental sulphur. Some, like primary phosphines (CH_3PH_2, $PhPH_2$, etc.), are spontaneously inflammable in air, while others, including triarylphosphines and aminophosphines, react only slowly at room temperature. These reactions are mainly free radical in mechanism and will be discussed more fully in Chapter 6. However, it is important to appreciate that this oxidation is one of the most obvious properties of phosphorus(III) compounds.

2.1.2 *Additions*

Phosphorus nucleophiles will attack polarized multiple bonds. Carbon–carbon multiple bonds require strongly electron-withdrawing substituents, for example ester or nitrile, to promote this attack; but isolated carbon–oxygen double bonds will often react by virtue of the polarity caused by electronegativity differences. Charged phosphorus nucleophiles, for example phosphide anions R_2P^-, will add to mildly polarized multiple bonds like styrene (2.33).

$$R_2P^- + Ph{-}CH{=}CH_2 \longrightarrow R_2P{-}CH_2{-}C^-H{-}Ph \quad (2.33)$$

(a) *Carbon–carbon bonds.* Although phosphines will not react with simple alkenes, they readily add to activated compounds like acrylonitrile, ethyl acrylate, etc. The initial nucleophilic addition is followed by a proton transfer to give a phosphorus ylid (2.34). However, this reaction appears to be reversible

$$R_3P\!: + CH_2{=}CHY \rightleftharpoons R_3P^+{-}CH_2{-}C^-H{-}Y \longrightarrow R_3P^+{-}C^-H{-}CH_2{-}Y \quad (2.34)$$

$Y = {-}CN, {-}COOR, {-}CONH_2$, etc.

(with the equilibrium well to the *left*) and, unless it is carried out in the presence of an acid or an aldehyde, which reacts with the ylid as it is formed, polymeric material results. Similarly activated triple bonds undergo an analogous reaction.

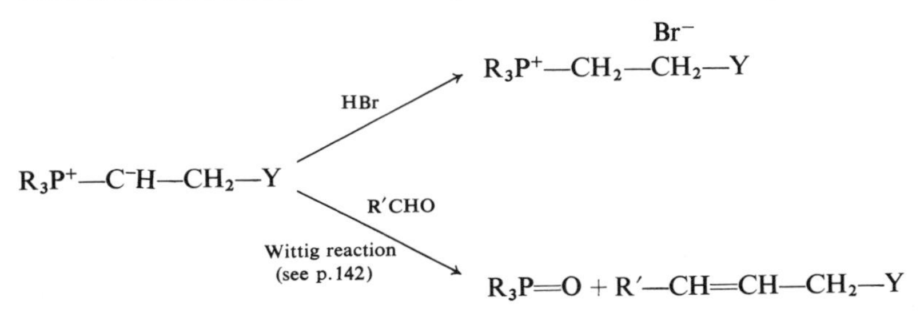

$$\text{p-benzoquinone} + :PPh_3 \longrightarrow \left[\text{zwitterion: } O^-,\ P^+Ph_3,\ H,\ H \right] \longrightarrow \text{ylid: } \overset{-}{C}\text{—}P^+Ph_3,\ OH \longleftrightarrow \text{phenoxide: } O^-,\ P^+Ph_3,\ OH$$

(2.34)

In some cases stable adducts are formed without the addition of acid. *p*-Benzoquinone reacts with triphenylphosphine to give the resonance-stabilized ylid (2.34), which is a crystalline solid stable to above 300 °C.

$$CH_3OOC\text{—}C\equiv C\text{—}COOCH_3 \xrightarrow{Ph_3P:} \left[Ph_3P^+\text{—}C(COOCH_3)\text{=}C^-\text{—}COOCH_3 \right]$$

(2.35) (2.37)

$$(2.37) \xrightarrow{SO_2/H_2O} Ph_3P^+\text{—}C(COOCH_3)\text{=}C(SO_3^-)\text{—}COOCH_3$$

$$(2.37) \xrightarrow{CO_2} Ph_3P^+\text{—}C(COOCH_3)\text{=}C(CO_2^-)\text{—}COOCH_3$$

$$(2.37) \xrightarrow{Ph_3P:} Ph_3P^+\text{—}C^-(COOCH_3)\text{—}C^-(COOCH_3)\text{—}P^+Ph_3$$

(2.36)

Dimethyl acetylenedicarboxylate (2.35) reacts with two molecules of triphenylphosphine to give, among other products, the diylid (2.36). That the reaction takes place via the intermediate (2.37) has been confirmed by trapping with sulphur dioxide and carbon dioxide. Presumably the negative charge in the initial addition product (2.37) is sufficiently delocalized to allow nucleophilic attack at a carbanion.

Primary and secondary phosphines (RPH_2 and R_2PH) also react with multiple bonds to give addition products (2.38), but the mechanism of these reactions is more obscure and in many cases undoubtedly involves radicals (see p. 214).

$$Ph_2PH + H_2C{=}CHR \longrightarrow Ph_2P{-}CH_2{-}CH_2{-}R$$

(2.38)

Phosphites will add to multiple bonds in a similar way to phosphines, but the final products are often different owing to the possibility of further reactions.

Trialkyl phosphites react with cyclopentenone, in the presence of a suitable proton donor like a phenol or an alcohol, to give intermediate salts (2.39). Unlike the analogous compound derived from the addition of a phosphine, this salt is unstable and undergoes an Arbusov-type reaction to give a phosphonate (2.40).

O

:P(OR)$_3$

O$^-$

P$^+$(OR)$_3$

(2.39)

R′OH

OH

P$^+$(OR)$_2$

O—R —OR′

O

O

P(OR)$_2$

+ R′OR

(2.40)

A similar reaction takes place with crotonaldehyde, in the presence of alcohols or phenols, to give the phosphonate acetal (2.42). In the absence of proton donors, these reactions go in very low yield, probably because a number of alternative pathways are open to the intermediates (2.39) and (2.41).

$$(RO)_3P\colon + CH_3{-}CH{=}CH.CHO \longrightarrow (RO)_3P^+{-}CH(CH_3){-}CH{=}CH{-}O^- \quad (2.41)$$

$$\xrightarrow{R'OH} (RO)_2P^+(O{-}R)\!-\!CH(CH_3){-}CH_2{-}CHO \;\; [^-OR'] \longrightarrow (RO)_2P(=O){-}CH(CH_3){-}CH_2{-}CHO$$

$$\xrightarrow{2R'OH} (RO)_2P(=O){-}CH(CH_3){-}CH_2{-}CH(OR')_2 \quad (2.42)$$

Trivalent-phosphorus esters also react with α,β-unsaturated acids at the alkene double bond, the initial addition being followed by an Arbusov reaction, either intramolecularly or intermolecularly, to give the phosphonate (2.43).

$$R'P(OR)_2 + CH_2{=}CH.COOH \longrightarrow R'P^+(OR)_2{-}CH_2{-}C^-H{-}COOH \longrightarrow \text{cyclic } (RO)(R')P^+{-}CH_2{-}CH_2{-}C(=O){-}O^- \;[O{-}R]$$

$$\longrightarrow R'P(=O)(OR){-}CH_2{-}CH_2{-}CO{-}OR \quad (2.43)$$

With allyl phosphites, intramolecular additions leading to rearrangement are possible. The product (2.44) from heating triallyl phosphite could be the result of a simple phosphite rearrangement (see p. 55), but reactions with substituted allyl phosphites (2.45) show that the reaction goes via a cyclic transition state. A similar reaction takes place with propargyl phosphites (2.46), leading to allenyl phosphonates (2.47).

$$(CH_2{=}CH{-}CH_2{-}O)_3P \xrightarrow{150\,°C} (CH_2{=}CH{-}CH_2{-}O)_2\overset{O}{\overset{\|}{P}}{-}CH_2{-}CH{=}CH_2$$

(2.44)

$$(RO)_2P{:}\ \text{[O–CH(CH}_3\text{)–CH=CH}_2\text{]} \longrightarrow (RO)_2P(=O){-}CH_2{-}CH{=}CH{-}CH_3$$

(2.45)

$$(RO)_2P{:}\ \text{[O–CH}_2\text{–C}{\equiv}\text{C–H]} \longrightarrow (RO)_2P(=O){-}CH{=}C{=}CH_2$$

(2.46) (2.47)

(b) *Carbon–oxygen bonds.* Tertiary phosphines react only slowly with ketones and aldehydes, unless the carbonyl group is activated by electron-withdrawing substituents. The addition usually takes place by initial attack at the carbon atom to give the zwitterion (2.48) which, in the presence of acid, gives the

$$R_3P{:} + RHC{=}O \longrightarrow R_3P^+{-}C(O^-)(H)(R) \xrightarrow{HX} R_3P^+{-}C(OH)(H)(R)\ X^-$$

(2.48) (2.49)

α-hydroxyphosphonium salt (2.49). Primary and secondary phosphines react in a similar way, but are able to transfer a proton to give α-hydroxyphosphines (2.50) as final products.

$$R_2PH + R'C(=O){-}R \longrightarrow R_2P^+(H){-}C(O^-)(R')(R) \longrightarrow R_2P{-}C(OH)(R')(R)$$

(2.50)

Chlorophosphines (2.51; X = Cl) and aminophosphines (2.51; $X = NR_2$) undergo reaction much more readily, but the products (2.52) are quite different and obviously involve a rearrangement, the mechanism of which is far from clear.

$$R_2PX + R'CHO \longrightarrow R_2P(=O){-}CH(R'){-}X$$

(2.51) (2.52)

Alkyl phosphites react in a similar way to phosphines to give the zwitterion (2.53), but this can then undergo an Arbusov reaction to give an alkoxy-phosphonate (2.54).

$$(RO)_3P: + H{-}C(R'){=}O \longrightarrow (RO)_2\overset{O-R}{\overset{|}{P^+}}{-}\underset{\underset{O^-}{|}}{CH}{-}R' \longrightarrow (RO)_2\overset{O}{\overset{\|}{P}}{-}\underset{\underset{OR}{|}}{CH}{-}R'$$

(2.53) (2.54)

Phosphites can also react with α-haloketones by nucleophilic attack at the carbonyl group (the Perkov reaction). By analogy with phosphines (p. 57), phosphites might be expected to react with α-haloketones either with direct displacement of the halogen by attack at the carbon atom (2.55) or with displacement of the carbanion (2.56) by attack at the halogen atom. In fact phosphites react with α-haloketones by a completely different mechanism involving

$$(RO)_3P: + \overset{R}{\overset{|}{C}}H(Br){-}C(=O){-}CH_3 \longrightarrow (RO)_3P^+{-}\overset{R}{\overset{|}{C}}H{-}\overset{O}{\overset{\|}{C}}{-}CH_3 \quad Br^-$$

(2.55)

$$(RO)_3P: + Br{-}\overset{R}{\overset{|}{C}}H{-}C(=O){-}CH_3 \longrightarrow (RO)_3P^+{-}Br \quad RC^-H{-}\underset{\underset{O}{\|}}{C}{-}CH_3$$

(2.56)

the initial attack of phosphite at the carbonyl group. Whether this attack is at carbonyl oxygen or carbonyl carbon has been the subject of considerable debate, but the most widely accepted mechanism involves nucleophilic attack at carbonyl carbon (2.57) followed by rearrangement to give the intermediate salt (2.58), which decomposes by an Arbusov reaction to the vinyl phosphate (2.59).

$$R'{-}\overset{O}{\overset{\|}{C}}{-}CH_2Br + (RO)_3P: \longrightarrow \left[(RO)_3P^+{-}\overset{O^-}{\overset{|}{\underset{\underset{R'}{|}}{C}}}{-}CH_2Br\right] \longrightarrow (RO)_3P^+{-}O{-}\overset{R'}{\overset{|}{C}}{=}CH_2 \quad Br^-$$

(2.57) (2.58)

$$\longrightarrow (RO)_2\overset{O}{\overset{\|}{P}}{-}O{-}CR'{=}CH_2 + RBr$$

(2.59)

Well-authenticated examples of the attack of phosphorus nucleophiles on the oxygen of carbonyl groups are rare, except in the case of α-diketones (see p. 79). The attack then is probably to some extent concerted and the displaced negative charge is stabilized by the phosphorus. Other cases which probably involve initial attack at carbonyl oxygen also have a route for resonance stabilization of the displaced negative charge. Phthalic anhydride reacts readily with trialkyl phosphites to give biphthalyl (2.60) and trialkyl phosphate (2.61) in high yield. It seems reasonable to write this reaction as involving attack at an oxygen atom and elimination of phosphate to give a carbene, but recently it has been shown to involve the ylid (2.62), which presumably reacts with phthalic anhydride in a type of Wittig reaction (see p. 142).

+ $(RO)_3P$ ⟶ (2.62) $P(OR)_3$

(2.60) + $(RO)_3P{=}O$ (2.61)

(c) *Other multiple bonds*. Examples of attack on oxygen by trivalent phosphorus are more common with N=O bonds than with C=O bonds. Both nitroso and nitro compounds are readily deoxygenated by phosphites and, less readily, by phosphines. Considerable evidence has been accumulated that these deoxygenations lead to nitrene intermediates; for example, *o*-nitrobiphenyl reacts with phosphites to give carbazole (2.64) via the nitrene (2.63) (although other products are formed in the presence of base, see Table 15).

NO_2 $\xrightarrow{2(RO)_3P:}$ (2.63) ⟶ (2.64)

Table 15 Reactions of trivalent phosphorus involving addition to multiple bonds

$$Ph_3P: + 2(NC)_2C{=}C(CN)_2 \longrightarrow$$ (cyclic phosphorane: P(Ph)₃ in a five-membered ring bearing NC, CN groups on each carbon)

$$R_3P: + N{\equiv}N^+{-}C^-R_2 \longrightarrow R_3P{=}N{-}N{=}CR_2$$

phosphazine

$$MeP(=S)Cl_2 + n\text{-}Bu_3P: \longrightarrow MePCl_2 + n\text{-}Bu_3PS$$

$$Ph_2P^-Na^+ + PhCHO \longrightarrow Ph_2P{-}CHPh{-}O^-\ Na^+$$

$$\text{cyclohexane-1,2-dione} + (EtO)_3P \xrightarrow{120\,°C} \text{2-ethoxy-2-}[P(O)(OEt)_2]\text{cyclohexanone}$$

$$(MeO)_3P: + PhCO{-}CH{=}CH{-}COPh \longrightarrow (MeO)_3P{=}C(COPh){-}CH_2{-}CO{-}Ph$$

$$(RO)_3P: + CH_2{=}CH{-}COOH \longrightarrow (RO)_2P(=O){-}CH_2{-}CH_2{-}COOR$$

$$(EtO)_3P: + EtOOCN{=}NCOOEt \longrightarrow (EtO)_2P(=O){-}O{-}C(OEt){=}N{-}N(Et){-}COOEt$$

$$\text{2-nitrobiphenyl} \xrightarrow[Et_2NH]{(EtO)_2PMe} \text{3-H-3-phenyl-2-}NEt_2\text{-azepine} + \text{carbazole (N–H)}$$

13%   67%

$$\text{2-(2-nitrophenyl)benzotriazole (}O_2N\text{)} \xrightarrow{(EtO)_3P} \text{dibenzotetrazapentalene (}N^+, N^-\text{)}$$

Table 15 - continued

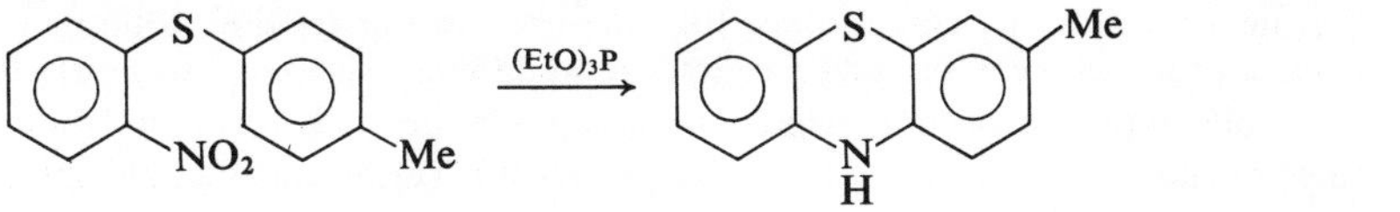

The generally accepted mechanism for these deoxygenations involves initial attack of trivalent phosphorus on oxygen; and the intermediacy of nitroso compounds in the deoxygenation of nitro groups has been demonstrated recently by the reaction of triethyl phosphite with nitrobenzoxazole (2.65). The product of this reaction (2.67) is thought to be formed by trapping the intermediate nitroso compound (2.66) as shown.

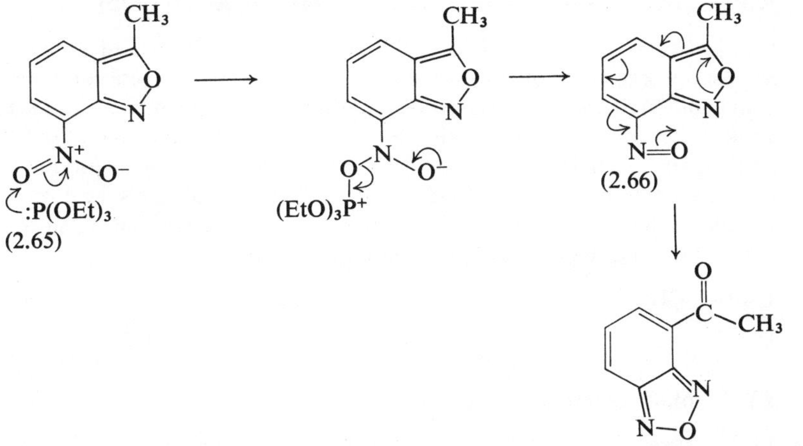

(2.65)

(2.66)

(2.67)

Trivalent-phosphorus compounds will react at the oxygen atom with highly polarized double bonds like those in amine oxides (2.68), sulphoxides (2.69) and even, in some cases, oxides of other phosphorus compounds (2.70). Other examples of phosphorus additions to multiple bonds can be found in Table 15.

$$R_3N^+O^- + R'_3P \longrightarrow R_3N + R'_3P{=}O \quad (2.68)$$

$$R_2S^+O^- + R'_3P \longrightarrow R_2S + R'_3P{=}O \quad (2.69)$$

$$Cl_3P{=}O + R'_3P \longrightarrow Cl_3P + R'_3P{=}O \quad (2.70)$$

2.1.3 *Electron donation*

Trivalent-phosphorus compounds, like amines, can donate their lone pair to form coordination compounds. The important difference is that phosphorus has available empty 3d orbitals which can accept charge back from any suitable filled orbitals on the acceptor atom (see pp. 32–40). This means that phosphines and phosphites form much stronger coordinate links than amines.

Most examples of this type of behaviour are inorganic. Phosphines and phosphites have a well-known ability to donate electrons to transition metals and metal ions, and are among the most-common ligands in metal-complex formation.

Trivalent-phosphorus compounds behave in a similar way to amines in that they form stable adducts with compounds of group IIIa elements. However, unlike the complexes with transition metals, these adducts with phosphines

$$R_3P: + MR'_3 \longrightarrow R_3P^+{-}M^-R'_3 \qquad (M = B, Al, Ga \text{ or } In)$$

tend to be less stable than those with amines. This has been used as evidence against the occurrence of any back donation (d_π bonding) in these compounds.

In addition to donor ability, trivalent-phosphorus compounds can also act as weak acceptors and form weakly bound adducts with donor molecules like amines; for example, the adduct (2.71) between phosphorus trichloride and trimethylamine has a P—N bond energy of 26·8 kJ mol^{-1} (6·4 kcal mol^{-1}). Details of the bonding in these compounds are not known, but possibly the empty low-lying 3d orbitals of phosphorus play a part.

$Cl_3P{-}NCH_3$
(2.71)

2.2 Electrophilic reactivity

Both nitrogen and phosphorus in the trivalent state have a lone pair of electrons and show nucleophilic character. In addition to this, phosphorus has fairly low energy empty d-orbitals which allow many of its compounds to show electrophilic character by stabilizing transition states, or intermediates, in reaction with nucleophiles. This is well illustrated by the different behaviour of nitrogen and phosphorus halides on hydrolysis. For phosphorus trichloride the involvement of d-orbitals leads to stabilization of the intermediate (2.72). In the absence of similar stabilization, nitrogen trichloride is hydrolysed with the initial attack on a halogen atom.

$$H_2\ddot{O}: \curvearrowright PCl_3 \longrightarrow \left[H_2O^+{-}P^-(Cl)_3\right] \longrightarrow H_2O^+{-}P(Cl)_2 + Cl^-$$

(2.72)

$$\xrightarrow{2H_2O} O{=}P(H)(OH)_2$$

This is a much-simplified picture of the hydrolysis of phosphorus trichloride.

$$HOCl + H_3O^+$$

$$\uparrow H_2O$$

$$H_2\ddot{O}: \; Cl\ddot{N}\begin{smallmatrix} Cl \\ Cl \end{smallmatrix} \longrightarrow H_2O^+{-}Cl + \underset{..}{N^-}\begin{smallmatrix} Cl \\ Cl \end{smallmatrix} \xrightarrow{H^+} H\ddot{N}\begin{smallmatrix} Cl \\ Cl \end{smallmatrix}$$

$$\downarrow 2H_2O$$

$$H_3N + 2HOCl$$

Having established the feasibility of electrophilic character, we must consider which trivalent-phosphorus compounds are likely to show it. The bonding of phosphorus to a more electronegative element will enhance this character. Firstly, because it will set up a favourable dipole (2.73) (this will be more

$$>\overset{\delta+}{P}{-}\overset{\delta-}{X}$$
(2.73)

difficult with the much more electronegative nitrogen) and, secondly, because a partial positive charge on the phosphorus atom will assist overlap with the 3d orbitals (see p. 31). An electrophilic reaction will be further encouraged by the presence of a good leaving group on the phosphorus atom. However, this criterion is not absolutely necessary because aminophosphines (2.74) show considerable electrophilic character and amide anion (R_2N^-) is not a good leaving group.

$R'_2P{-}NR_2$
(2.74)

Chlorophosphines (2.75) undergo electrophilic reaction even more readily, and chloride ion is a much better leaving group.

$R_2P{-}Cl$
(2.75)

Considerations of this sort suggest that halophosphines (2.75), aminophosphines (2.74) and compounds which contain oxygen–phosphorus bonds (phosphites (2.76), phosphinites (2.77), etc.) will probably show electrophilic

$(RO)_3P$ (2.76) $RO{-}PR'_2$ (2.77)

character. As we shall see in the following pages, this is the case, although compounds with oxygen–phosphorus bonds are complicated to some extent by the very high P=O bond energy.

2.2.1 *Neutral nucleophiles*

The reaction of phosphorus halides with water to give phosphorus acids and hydrogen halide is well known and has already been discussed. The reaction of alcohols and carboxylic acids is analogous, leading to the alkyl halide (2.78) and acyl halide (2.79) respectively.

$$PCl_3 + 3ROH \longrightarrow 3RCl + (HO)_2P(H){=}O$$

(2.78)

$$PCl_3 + 3R\,C({=}O){-}O{-}H \longrightarrow 3RC({=}O)Cl + (HO)_2P(H){=}O$$

(2.79)

The reaction of amines with trivalent-phosphorus halides is similar, but the products vary depending on the conditions. In the presence of excess primary or secondary amines, most halophosphines give aminophosphines (2.80); but as the amount of primary amine is reduced, complex products (e.g. 2.81) begin to form.

$$R_2PCl + 2PhNH_2 \longrightarrow R_2P{-}NH{-}Ph$$

(2.80)

$$PCl_3 + MeNH_2 \longrightarrow P_4(NMe)_6$$

(2.81)

2.2.2 *Organo-metallic nucleophiles*

There are many examples of reactions between trivalent-phosphorus compounds and metal alkyls, or aryls; most of these can be interpreted in terms of nucleophilic attack by the organic residue on the phosphorus atom. Although these reactions are some of the most important in the synthesis of organophosphorus compounds, they are sufficiently similar for a minimum of examples to demonstrate their utility.

The most commonly met reaction of this type is that of an alkyl, or aryl, Grignard reagent with a halophosphine. This is one of the most convenient routes to phosphines (2.82).

$$PCl_3 + 3\,ArMgX \longrightarrow Ar_3P$$

(2.82)

Similar reactions take place between other metal alkyls (e.g. 2.83) and both phosphines and phosphorus esters.

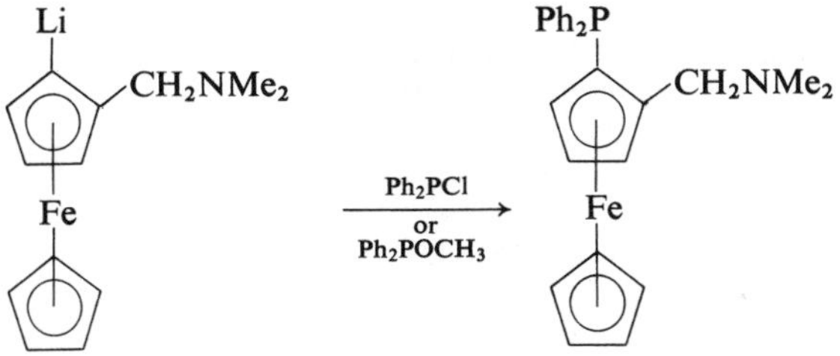

Li, CH_2NMe_2, Fe (2.83) $\xrightarrow[\text{or Ph}_2\text{POCH}_3]{\text{Ph}_2\text{PCl}}$ Ph_2P, CH_2NMe_2, Fe

(2.83)

2.2.3 *π-Bonded systems*

Phosphorus trichloride reacts readily with aromatic hydrocarbons in the presence of aluminium chloride (or other Lewis acids). This reaction appears to go by the normal Friedel–Crafts reaction, involving attack by electrophilic phosphorus (2.84).

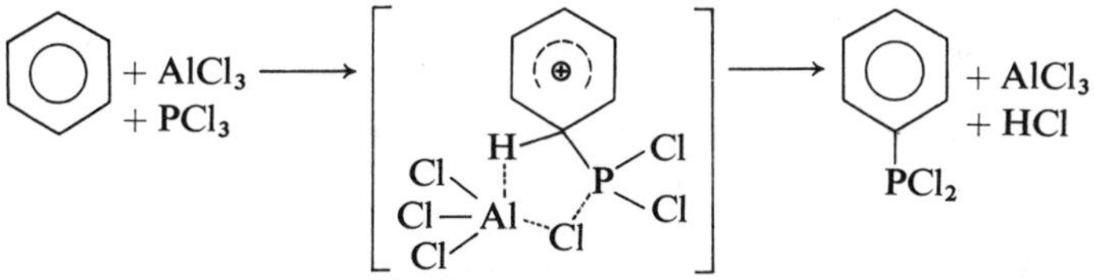

$C_6H_6 + AlCl_3 + PCl_3 \longrightarrow$ [H, Cl, P, Cl, Cl—Al—Cl, Cl, Cl, ⊕] $\longrightarrow C_6H_5PCl_2 + AlCl_3 + HCl$

(2.84)

Examples of reactions involving unsaturated aliphatic systems are also known (e.g. the formation of 2.85). Those involving conjugated dienes are probably

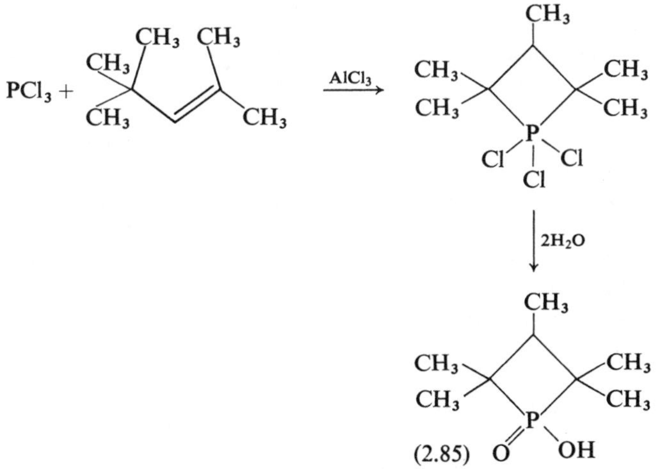

$PCl_3 +$ CH_3 CH_3 CH_3 CH_3 CH_3 $\xrightarrow{AlCl_3}$ CH_3, CH_3, CH_3, CH_3, CH_3, P, Cl, Cl, Cl

$\downarrow 2H_2O$

CH_3, CH_3, CH_3, CH_3, CH_3, P, O, OH

(2.85)

concerted cycloadditions (see below), while in other cases it is often difficult to distinguish between radical and ionic pathways.

Table 16 Trivalent phosphorus as an electrophilic reagent

Reactants	Conditions	Products
$PCl_3 + 3Me_3C{-}C{\equiv}C{-}Li$	$\xrightarrow{THF}$	$(Me_3C{-}C{\equiv}C{-})_3P{:}$ 98%
$EtPBr_2 + 2Et_3SnCH_2.COOEt$	$\longrightarrow$	$EtP(CH_2.COOEt)_2$ 66%
$PhPCl_2 + 2NaNCO$	$\xrightarrow[reflux]{MeCN}$	$PhP(NCO)_2$ 67%
$EtOPCl_2 + 2n\text{-}BuMgCl$	$\xrightarrow{ether}$	$EtOP(n\text{-}Bu)_2$ 53% b.p. 88–92 °C at 12 mm
$(R_2N)_3P + R{-}C(OR){=}NH$	$\longrightarrow$	$R{-}C(OR){=}N{-}P(NR_2)_2$
$(PhO)_3P + 3BuO^-Na^+$	$\xrightarrow{BuOH}$	$(BuO)_3P + 3PhOH$ 80% b.p. 122–123 °C at 12 mm

2.3 Dienophilic reactivity

In addition to showing nucleophilic and electrophilic reactivity, certain trivalent phosphorus compounds can react as dienophiles. This 1,4-cycloaddition reaction of phosphorus compounds with dienes is an important route to phosphorus heterocycles. As well as additions to conjugated dienes, phosphorus compounds can show dienophilic reactivity towards α,β-unsaturated ketone (C=C—C=O) and α-diketone (O=C—C=O) systems. Mechanistically these reactions can be divided into two groups, those where the initially formed adduct is stable and those in which it can undergo further transformation under the reaction conditions. Which of these mechanisms operates depends on the substituents present on the phosphorus compound. The criteria for initial addition of the phosphorus compound appear to be the same as those for the Diels–Alder reaction, and so it seems that the first step in both mechanisms involves a concerted addition of trivalent phosphorus to the diene. Examples of trivalent phosphorus acting as a dienophile are given in Table 17.

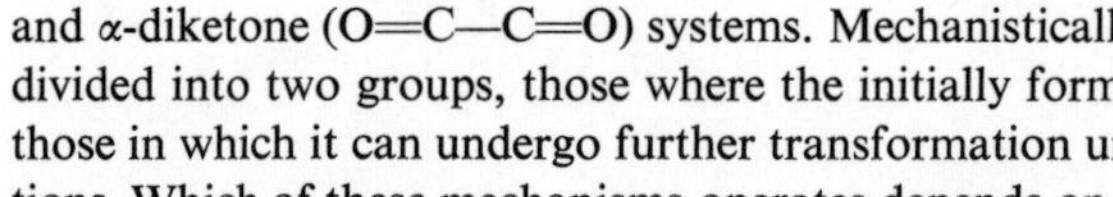

Table 17 Trivalent-phosphorus compounds as dienophiles

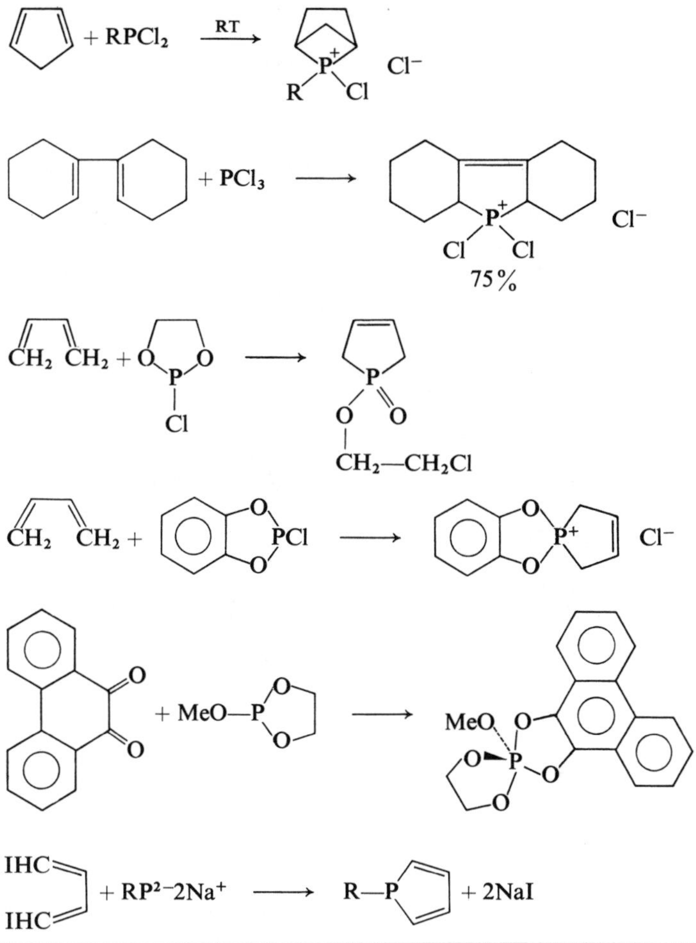

2.3.1 *Stable adduct formation*

Trivalent-phosphorus halides react with dienes to form cyclic adducts. For example, phosphorus trichloride reacts with butadiene to form the cyclic phosphonium salt (2.86). (This compound could equally well be represented by the pentacovalent form (2.87).)

PCl_3 + [diene] ⟶ Cl_2P^+ [ring] Cl^- (2.86) Cl_3P [ring] (2.87)

These adducts are perfectly stable under the reaction conditions. However, they will react with hydroxylic compounds (e.g. water or alcohols) to give acids (2.88) or esters (2.89), presumably by an initial nucleophilic attack on phosphorus

(2.88)

(2.89)

followed by an Arbusov reaction. In some reactions of this type, double-bond isomerization to the 2-position (2.90) appears to take place.

(2.90)

Both alkyl and aryl phosphorus dihalides undergo analogous reactions to form stable adducts (2.91) which can be converted into the corresponding phosphine oxide (2.92) on treatment with water or alcohol.

$RPCl_2$ + [diene] ⟶ (2.91) ⟶ (2.92)

A few examples of non-conjugated dienes reacting with chlorophosphines are known. Bicyclo[2.2.1]heptadiene (2.93) and methyldichlorophosphine react to give the adduct (2.94). Presumably the addition is assisted by the favourable geometry of the system.

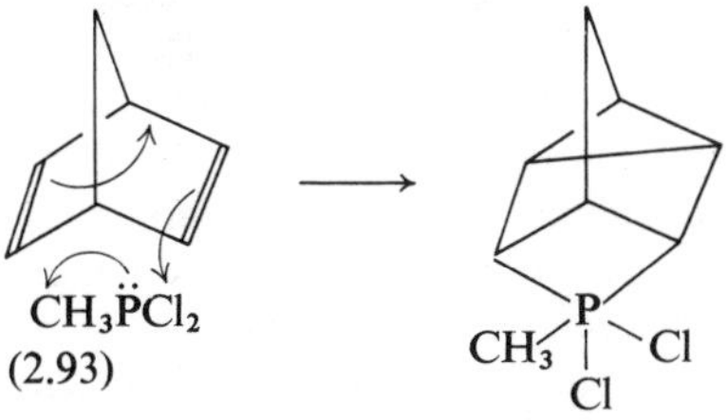

$CH_3\ddot{P}Cl_2$
(2.93)

CH_3 P Cl Cl
(2.94)

Most phosphorus esters undergo further reaction after the initial addition and will be discussed in the next section. However, as we have already seen (p. 55), aromatic phosphites $(ArO)_3P$, phosphinites $ArOPR_2$, etc. will only undergo the Arbusov reaction at high temperatures, and this has allowed a number of stable adducts (e.g. 2.95) to be isolated using aryl esters.

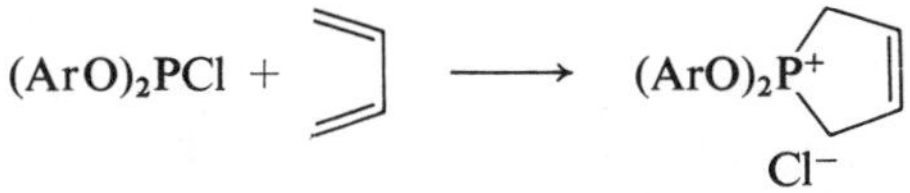

$(ArO)_2PCl$ + ⟶ $(ArO)_2P^+$ Cl^-
(2.95)

One reaction which has been extensively studied in recent years is that between phosphites and α-diketones. The product is the simple adduct (2.96), which is usually stable.

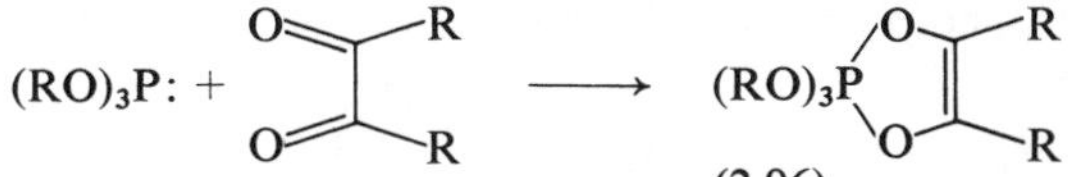

$(RO)_3P$: + ⟶ $(RO)_3P$
(2.96)

There has been considerable discussion about the structure of these adducts, both a covalent form (2.97) and a number of dipolar forms (2.98, 2.99 and 2.100) are possible. However, the physical properties of the adducts (solubility in non-

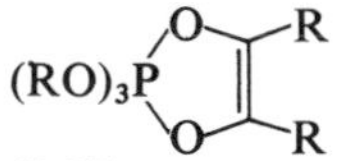

$(RO)_3P$
(2.97)

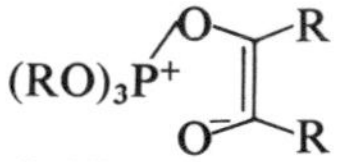

$(RO)_3P^+$
(2.98)

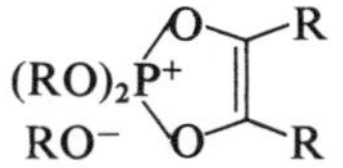

$(RO)_2P^+$ RO^-
(2.99)

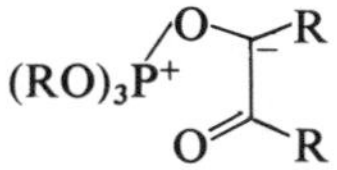

$(RO)_3P^+$
(2.100)

polar solvents, low dipole moments, etc.) suggest at least a predominance of the covalent form, and spectroscopic techniques have recently been used to confirm the covalent structure. Their infrared spectra show no carbonyl absorption and their proton magnetic resonance spectra, although confused by pseudo-rotation (p. 31) and exchange, are best interpreted in terms of the covalent structure (2.97). Final confirmation was obtained from the phosphorus-31 nuclear magnetic resonance spectra of the adducts in which the chemical shift of the phosphorus resonance is close to that for covalent phosphorus(V) compounds and far removed from that normally associated with phosphonium salt structures.

Although the compounds appear to be covalent in the ground state, dipolar forms may contribute as reaction intermediates. For example, they readily undergo ring opening with nucleophilic reagents like water and alcohols to give phosphates (2.101).

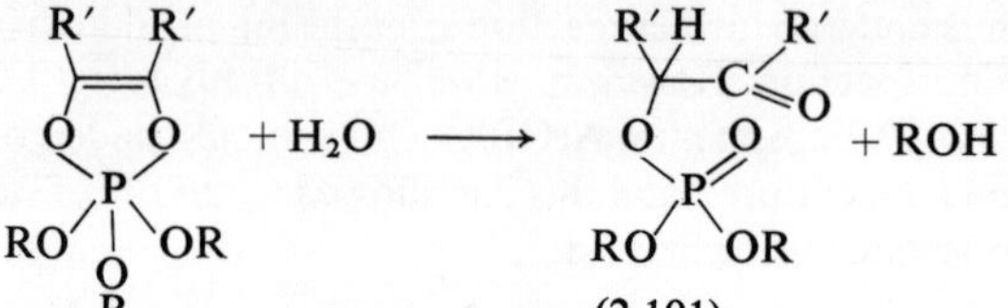

R′ R′ ; O O ; P ; RO OR ; O ; R $+ H_2O \longrightarrow$ R′H R′ ; C=O ; O O ; P ; RO OR $+$ ROH

(2.101)

Analogous adducts of aminophosphines $(R_2N)_3P$, phosphonites $(RO)_2PR$ and phosphinites $ROPR_2$ have also been prepared and, for these, dipolar structures seem likely in some instances.

α,β-Unsaturated ketones react with phosphorus esters by one of two routes. 3-Benzylidene-2,4-pentanedione (2.102) and trialkyl phosphites give adducts (2.103) analogous to those obtained from α-diketones. However, dibenzoylethylene undergoes Michael addition to the olefinic double bond followed by proton transfer to give a phosphorus ylid (2.104).

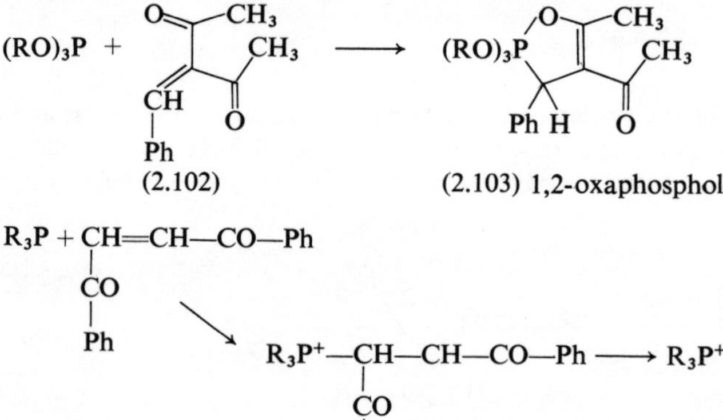

$(RO)_3P$ + O CH₃ CH₃ CH Ph O $\longrightarrow$ $(RO)_3P$ O CH₃ CH₃ Ph H O

(2.102) (2.103) 1,2-oxaphospholen

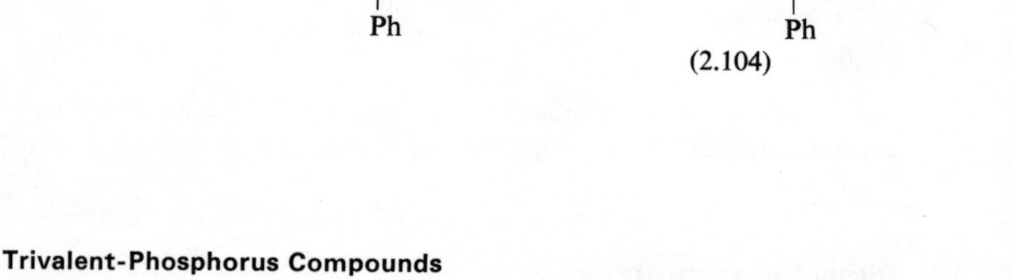

R_3P + CH=CH—CO—Ph (CO—Ph on first CH)

$\searrow$ R_3P^+—CH(COPh)—CH—CO—Ph $\longrightarrow$ R_3P^+—C^-(C=O Ph)—CH_2—C(=O)—Ph

(2.104)

Carbenes, which can be thought of as lower homologues of dienes, react readily with phosphines to give ylids (2.105). This reaction is reversible on ultraviolet irradiation.

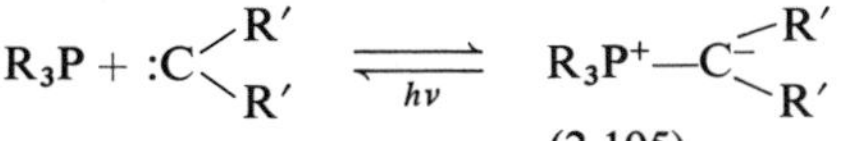

$$R_3P + :C\begin{smallmatrix}R'\\R'\end{smallmatrix} \underset{h\nu}{\rightleftharpoons} R_3P^+{-}C^-\begin{smallmatrix}R'\\R'\end{smallmatrix}$$

(2.105)

2.3.2 *Adducts which undergo further reaction*

Both dichlorophosphinites Cl_2POR and chlorophosphonites $ClP(OR)_2$ react readily with dienes presumably to form the adducts (2.106a and b). However, this product is able to undergo further reaction through attack by a chloride ion to give the Arbusov products (2.107a and b).

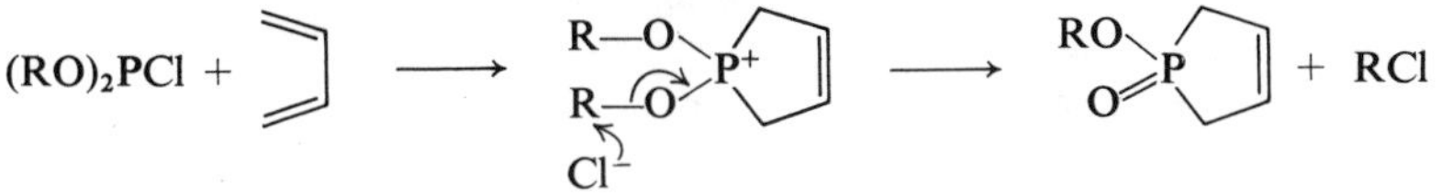

$(RO)_2PCl$ + diene ⟶ (2.106a) ⟶ (2.107a) + RCl

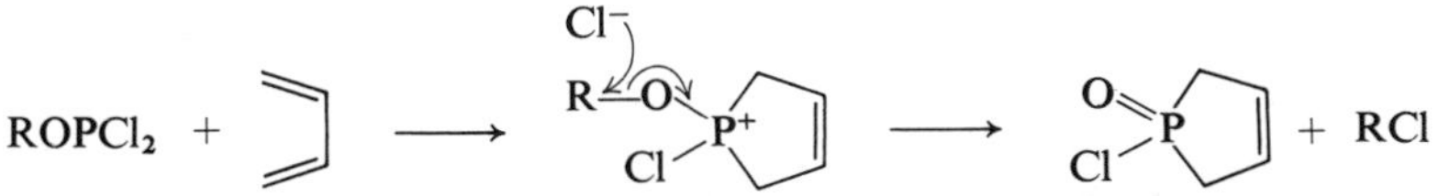

$ROPCl_2$ + diene ⟶ (2.106b) ⟶ (2.107b) + RCl

The analogous aryl phosphorus compounds undergo a similar initial reaction but, because the Arbusov reaction requires forcing conditions with aryl–oxygen bonds (p. 55), these adducts (2.108) are stable.

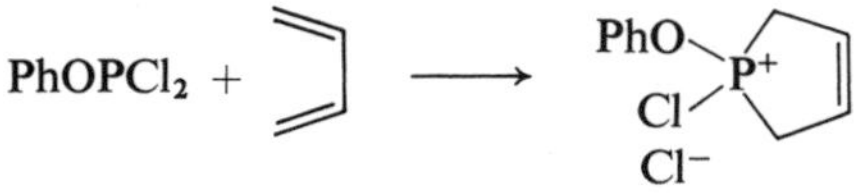

$PhOPCl_2$ + diene ⟶ (2.108)

Phenyldichlorophosphine (2.109) reacts with 1,4-diphenylbutadiene in an apparently abnormal manner to give the phosphole (2.110), which contains the phosphorus analogue of the pyrrole nucleus. It has been suggested that this reaction actually proceeds via the normal adduct (2.111), followed by loss of hydrogen chloride, but no evidence exists for this.

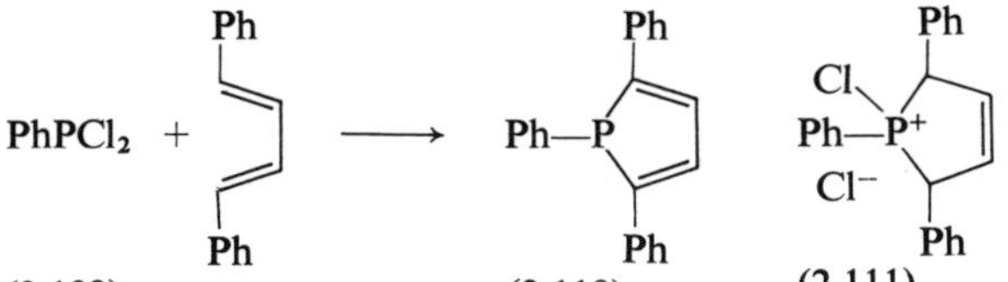

$PhPCl_2$ + 1,4-diphenylbutadiene ⟶ (2.110)

(2.109) (2.110) (2.111)

Problems

2.1 Name the following compounds.

Me_2PPh

$MeOPPh_2$

$(MeO)_2PEt$

$(PhO)_2POMe$

Ph_2PNMe_2

[See Table 11, p. 49.]

2.2 While nucleophiles, like amines and hydroxides (first row nucleophiles), usually attack carbon centres, trivalent-phosphorus compounds act as nucleophiles by attack at a variety of centres, e.g. halogen, oxygen, sulphur. Explain.
[See Chapter 2, pp. 51–63, and Chapter 1, p. 25.]

2.3 Trivalent-phosphorus nucleophiles can react with α-haloketones by an initial attack of the phosphorus atom at any one of four centres in the ketone. What are these four centres and how would you obtain information about the centre of initial attack on any particular haloketone?
[Kirby and Warren (1967), pp. 117–31. See also Chapter 2, pp. 57–61.]

2.4 Comment on the fact that triphenylphosphine reduces *erythro-N,N*-diethyl-dibromocinnamamide to *trans*-cinnamamide, and *meso*-dibromosuccinic acid to fumaric acid.

$$PhCHBr.CHBr.CONEt_2 + Ph_3P \longrightarrow (Ph)(H)C{=}C(H)(CONEt_2) + Ph_3PBr_2$$

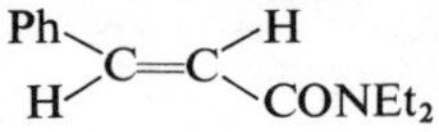

erythro

$$HOOC.CHBr.CHBr.COOH + Ph_3P \longrightarrow (H)(HOOC)C{=}C(COOH)(H) + Ph_3PBr_2$$

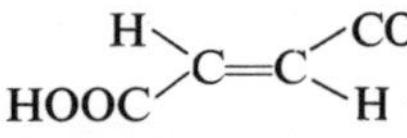

meso

[Hoffmann and Diehr (1962); Speziale and Tung (1963).]

2.5 Write mechanisms for the reactions given in Tables 14 and 15.

2.6 Suggest reasonable mechanisms for the following reactions.

(a) $2Ph_2C{=}C{=}O + (EtO)_3P \xrightarrow{RT}$ 2:1 adduct

$\downarrow$ heat 215 °C

$Ph{-}C{\equiv}C{-}Ph + (EtO)_3PO$

[Mukaiyama, Nambu and Okamoto (1962).]

(b) $Ph_2P—O—CHMe—CH{=}CH_2 + MeI$

$$\downarrow$$

$$Ph_2\overset{\overset{O}{\|}}{P}Me + Me—CHI—CH{=}CH_2 + MeCH{=}CH.CH_2I$$

[Kirby and Warren (1967), p. 41.]

(c) $Ph_3P + \begin{matrix} BrCH.COOMe \\ | \\ BrCH.COOMe \end{matrix} \xrightarrow{MeOH} Ph_3PO + MeBr + Ph_3P^+\underset{\underset{COOMe}{|}}{CH}—CH_2—COOMe\ \ Br^-$

$$\downarrow CHCl_3$$

$$Ph_3PBr_2 + MeOOC.CH{=}CH.COOMe$$

[Shaw and Tebby (1970).]

(d) $\underset{\text{(cyclic } OCH_2CH_2O\text{)}}{PCl} + Cl_2 \xrightarrow{-15\,^\circ C} Cl—CH_2CH_2O\overset{\overset{O}{\|}}{P}Cl_2$

[Griffith and Grayson (1965), p. 141.]

(e) $(RO)_3P \xrightarrow{R'.CH{=}CH.CH_2Cl} R'CH{=}CH.CH_2\overset{\overset{O}{\|}}{P}(OR)_2$

$(RO)_3P \xrightarrow{R'—\underset{\underset{Cl}{|}}{CH}—CH{=}CH_2} R'CH{=}CH.CH_2—\overset{\overset{O}{\|}}{P}(OR)_2$

[Griffith and Grayson (1964), p. 65.]

(f) $(CF_3)_2CO + R_2PH \xrightarrow{O_2} (CF_3)_2CH—O—\overset{\overset{O}{\|}}{P}R_2$

[Stockel (1968). See also Trippett (1971), pp. 29-40.]

(g) $Me_2C{=}CH—CH(S—SEt)Me + Ph_3P$

$$\downarrow$$

$$Me_2\underset{\underset{SEt}{|}}{C}—CH{=}CH—Me + Ph_3PS$$

[Moore and Trego (1962).]

Chapter 3
The Pentavalent State – Compounds Derived from PX_5

3.1 Pentacoordinate phosphorus(V) compounds (R_5P)

These compounds contain five single σ-bonds linking five separate groups and, when covalently bonded, use phosphorus 3d orbitals in their bonding. The trigonal-bipyramidal disposition (3.1) of these groups has been demonstrated in a number of compounds. However their stereochemistry is complicated by pseudo-rotation (see pp. 31–2).

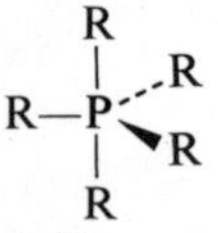

R
R—P R R
R

(3.1)

In compounds like (3.1), the phosphorus atom has no available unbonded electrons, and so its reactions are almost entirely those of an electrophile. Nucleophilic substitution at the phosphorus atom takes place fairly readily depending on the groups already coordinated to the central atom. When these groups include an atom which is both electronegative and a good leaving group (e.g. halide; but see also p. 72) the substitution is usually easy. Compounds which contain five carbon–phosphorus bonds, for example (3.2), are much more resistant to nucleophilic attack, although six-coordinate compounds (3.3) have been isolated.

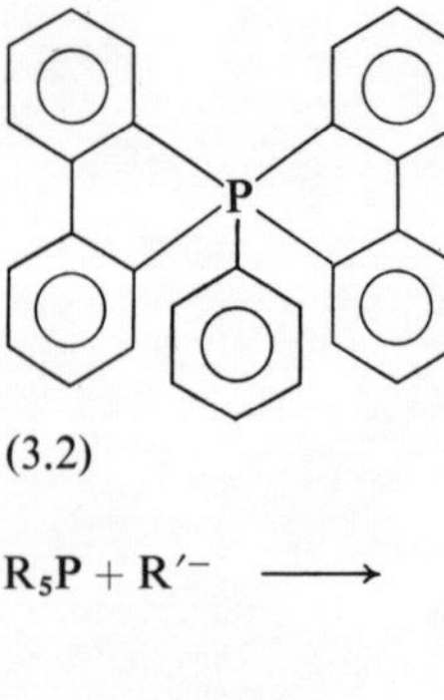

P

(3.2)

$$R_5P + R'^- \longrightarrow R_5P^-\text{—}R'$$

(3.3)

Phosphorus mechanisms of all types have been much less studied than those involving reaction at a carbon atom, and much less information is available. Whether substitutions at phosphorus are S_N1 or S_N2 (i.e. whether they proceed by mechanism **A** or mechanism **B**) is less clear than for reactions at a carbon atom, and a third mechanism **C** exists involving an intermediate (3.4c) (see above). This cannot happen with carbon, as no suitable orbitals are available. However,

$$R_4PX \longrightarrow \underset{(3.4a)}{\underset{X^-}{R_4P^+}} \xrightarrow{Y^-} R_4PY \qquad \textbf{A}$$

$$Y^- + \overset{R_4}{PX} \longrightarrow Y\text{---}\overset{R_4}{P}\text{---}X \longrightarrow YPR_4 + X^- \qquad \textbf{B}$$

$$Y^- + \overset{R_4}{PX} \longrightarrow \underset{(3.4c)}{\left[Y\text{—}\overset{R_4}{P}\text{—}X\right]^-} \longrightarrow YPR_4 + X^- \qquad \textbf{C}$$

there are many examples of pentacovalent-phosphorus compounds which are largely ionized in their stable state (see phosphonium salts, p. 127), and so possess structure (3.4a) without the presence of either a solvent or a reagent; this suggests that the S_N1 mechanism **A** could be important in many cases.

Another factor in the reaction of these compounds is the large gain in energy in forming a P═O bond (bond strength 545 kJ mol^{-1} (130 kcal mol^{-1}), see p. 34) or a P═N bond (bond strength 460 *k*J mol^{-1} (110 kcal mol^{-1}). Because of the tendency toward the formation of these two stable bonds, many of the initial products of nucleophilic substitution are unstable.

3.1.1 *Phosphorus(V) halides*

These are compounds which have at least one halogen atom bonded to the phosphorus atom and their structures range from ionic to covalent, depending on the nature of both the halogens and the other substituents. Bonding can even vary with the physical state; phosphorus pentachloride, for example, is ionic in the solid but covalent in the liquid state. However, compounds containing four alkyl or aryl substituents ($R_4P^+X^-$) are largely ionic and will be discussed later. Both covalent and ionic phosphorus(V) halides show reactivity towards a large range of nucleophilic reagents.

(a) *Hydroxylic nucleophiles.* Phosphorus halides react with water and alcohols to form phosphorus acids and their derivatives. This aspect of phosphorus chemistry can cause much confusion because of the apparent complexity of the phosphorus acids and their nomenclature.

An attempt to summarize the formation of these acids, based on the hydrolysis reactions of the phosphorus halides, has been made in Table 21.

The initial reaction of any pentavalent-phosphorus halide with water is

$$R_3P\begin{smallmatrix}X\\X\end{smallmatrix} + H_2O \longrightarrow R_3P\begin{smallmatrix}O^+H_2\\X\end{smallmatrix}\;X^- \longrightarrow R_3P\begin{smallmatrix}O{-}H\\X\end{smallmatrix} \qquad (3.5)$$

nucleophilic substitution of one halogen atom (3.5). The intermediate product can then form a P=O bond by elimination of a hydrogen halide; it does this rapidly because of the strength of the P=O bond. If all the substituents (R) are alkyl or aryl, the reaction stops at this stage and the product formed (3.6) is a

$$R_3P\begin{smallmatrix}O{-}H\\X\end{smallmatrix} \longrightarrow R_3P{=}O \qquad (3.6)$$

phosphine oxide. However, if the product of the initial substitution has other halide substituents, the substitution continues, forming various phosphorus acids (3.7, 3.8 and 3.9).

$$R_2PCl_3 \xrightarrow{H_2O} R_2P\begin{smallmatrix}Cl\\=O\end{smallmatrix} \xrightarrow{H_2O} R_2P\begin{smallmatrix}O{-}H\\=O\end{smallmatrix} \qquad (3.7)$$

dialkyl phosphinic acid

$$RPCl_4 \xrightarrow{H_2O} RP\begin{smallmatrix}Cl\\Cl\\=O\end{smallmatrix} \xrightarrow{2H_2O} RP\begin{smallmatrix}O{-}H\\O{-}H\\=O\end{smallmatrix} \qquad (3.8)$$

alkyl phosphonic acid

$$PCl_5 \xrightarrow{H_2O} \begin{smallmatrix}Cl\\Cl\\Cl\end{smallmatrix}P{=}O \xrightarrow{3H_2O} \begin{smallmatrix}H{-}O\\H{-}O\\H{-}O\end{smallmatrix}P{=}O \qquad (3.9)$$

phosphoric acid

The second stage of the hydrolysis appears analogous to that of acyl halides and will be discussed further on p. 107.

A similar nucleophilic substitution takes place with alcohols. The initial

$$R_3PCl_2 + R'OH \longrightarrow R_3P\begin{smallmatrix}OR'\\Cl\end{smallmatrix} \rightleftharpoons R_3P^+{-}O{-}R'\;\;Cl^- \qquad (3.10)$$

$$\downarrow$$

$$R_3PO + R'Cl$$

product in its ionic form (3.10) is identical with the intermediate in the Arbusov reaction (p. 54) and decomposes by the same pathway. This is a most effective way of converting alcohols into the corresponding alkyl halides with inversion of configuration.

If the phosphorus compound contains more halogen atoms (3.11, 3.12 or 3.13), the products are esters of phosphorus acids formed in a similar way to (3.7, 3.8 and 3.9). It is also possible to prepare the sulphur analogues of all these compounds but, because of the weaker nucleophilicity of the thiols, the substitutions are more difficult to perform.

$$R_2PCl_3 + 2R'OH \longrightarrow R_2P(OR')(=O) + R'Cl + 2HCl \quad (3.11)$$

alkyl dialkylphosphinate

$$RPCl_4 + 3R'OH \longrightarrow RP(OR')_2(=O) + R'Cl + 3HCl \quad (3.12)$$

dialkyl alkylphosphonate

$$PCl_5 + 4R'OH \longrightarrow (R'O)_3P{=}O + R'Cl + 4HCl \quad (3.13)$$

trialkyl phosphate

Other oxygen nucleophiles will carry out similar substitutions. The reaction of carboxylic acids with phosphorus pentahalides is a common method of preparation of acid halides and probably involves an intermediate like (3.14).

$$PCl_5 + R{-}C({=}O){-}OH \longrightarrow Cl_3P(Cl){-}O{-}C({=}O){-}R \;\; Cl^- \quad (3.14)$$

$$\downarrow$$

$$Cl_3P{=}O + Cl{-}C({=}O){-}R$$

Table 18 lists some examples of reactions of oxygen nucleophiles and sulphur nucleophiles with phosphorus pentahalides.

(b) *Nitrogen nucleophiles.* Amines and other nitrogen nucleophiles carry out substitutions at the pentavalent phosphorus atom, much as water does. However, the P=N bond is generally weaker than the P=O bond and this, together with

Table 18 The reaction of oxygen and sulphur nucleophiles with phosphorus(V) halides
(See also Table 21; p. 101)

$$Ph_3PCl_2 + 2PhOH \xrightarrow{2Et_3N} \underset{PhO^-}{Ph_3P^+{-}OPh} + 2EtN^+H \ \ Cl^-$$

$$Ph_3PCl_2 + H_2O \xrightarrow[\text{temp.}]{\text{room}} Ph_3P{=}O + 2HCl$$
m.p. 157 °C

$$(CF_3)_3PCl_2 + \begin{array}{c}COOH\\|\\COOH\end{array} \xrightarrow[\text{temp.}]{\text{room}} (CF_3)_3P{=}O + CO + CO_2 + 2HCl$$

$$PCl_5 + R_2P(=O){-}OH \xrightarrow[\text{reflux 1 h}]{\text{benzene}} R_2P(=O){-}Cl + Cl_3P{=}O + HCl$$
dialkylphosphonic chloride (R = Me; m.p. 67 °C)

$$Et_2PCl_3 + H_2O \longrightarrow Et_2P(=O)Cl + 2HCl$$
b.p. 102 °C at 15 mm

$$Ph_2PCl_3 + 2RSH \longrightarrow Ph_2P(=S){-}S{-}R + 2HCl + RCl$$
alkyl diphenylphosphinodithioate

$$RP(=O)(OR')_2 + 2PCl_5 \xrightarrow{80\text{–}90\,^\circ C} R{-}P(=O)Cl_2 + 2Cl_3P{=}O + 2R'Cl$$
alkylphosphonic dichloride (R = Me; m.p. 32 °C)

$$RPCl_4 + SO_2 \longrightarrow RP(=O)Cl_2 + SOCl_2$$

$$RPCl_4 + H_2S \longrightarrow RP(=S)Cl_2 + 2HCl$$
alkylphosphonothioic dichloride

Table 19 The reaction of nitrogen nucleophiles with phosphorus(V) halides

$$2PCl_5 + N_2H_4 \xrightarrow{100\,°C} Cl_3P{=}N{-}N{=}PCl_3 + 4HCl$$

$$3PCl_5 + H_2NP(=O)(OH){-}OH \longrightarrow Cl_3P{=}N{-}P(=O)Cl_2 + 4HCl + 2Cl_3P{=}O$$

N-[dichlorophosphinyl]-phosphorimidic trichloride

$$Ph_3PCl_2 + RNH_2 \xrightarrow[\text{benzene, } 5\,°C]{2Et_3N} Ph_3P{=}N{-}R + 2Et_3N^+H\ \ Cl^-$$

N-alkyl triphenylphosphine imide (R = Ph; m.p. 132 °C)

$$PCl_5 + PhNH_2 \xrightarrow[\text{reflux}]{CCl_4} Cl_3P{=}N{-}Ph + 2HCl$$

m.p. 180 °C

$$Ph_3PBr_2 + H_2N{-}N{=}CRR' \xrightarrow[\text{benzene, room temp.}]{2Et_3N} Ph_3P{=}N{-}N{=}CRR' + 2Et_3N^+H\ \ Br^-$$

(R=R′=Ph; m.p. 173 °C)

$$PCl_5 + H_2N{-}SO_2Ph \xrightarrow{150\,°C} Cl_3P{=}N{-}SO_2Ph + 2HCl$$

m.p. 54 °C

$$PhPCl_4 + H_2N{-}COOEt \xrightarrow[\text{temp.}]{Et_3N\text{, room}} PhP(Cl_2){=}N{-}COOEt + 2Et_3N^+H\ \ Cl^-$$

$$PCl_5 + R.CO.NH_2 \xrightarrow[\text{reflux}]{CCl_4} Cl_3P{=}N{-}CO{-}R + 2HCl$$

($R = CCl_3$; m.p. 77 °C)

$$(EtO)_3PCl_2 + PhSO_2NH_2 \xrightarrow[1\frac{1}{2}\,h]{160\,°C} (EtO)_3P{=}N{-}SO_2Ph + 2HCl$$

m.p. 85 °C

$$R_3PCl_2 + R'NH_2 \longrightarrow R_3P(Cl)\text{—}N(H)\text{—}R' + HCl$$

(3.15)

the lower electronegativity of nitrogen in comparison with oxygen, frequently enables the initial products of substitution (3.15) to be isolated. The analogous product (3.16), resulting from an attack by oxygen, always undergoes rapid elimination to form a phosphoryl compound (see p. 86). Since trivalent nitrogen can carry out further nucleophilic substitutions, the products may be complicated.

$$\left[R_3P(Cl)\text{—}O\text{—}H \right] \longrightarrow R_3P{=}O + HCl$$

(3.16)

Because the reactions between amines and pentavalent-phosphorus halides may yield such a variety and number of products, only the principles of these reactions will be discussed here. Detailed discussions can be found in reviews cited in the Bibliography.

The simplest of these reactions appears to be that between phosphorus pentachloride and ammonia. This reaction, and the products derived from it, have been the subject of a book, a dozen reviews, numerous patents and several thousand papers; the detailed mechanism of the reaction is extremely complex and is still not fully understood. However, the main features are similar to those already discussed.

Nucleophilic substitution by ammonia leads to the phosphine imide (3.17); this reacts with further pentachloride and ammonia molecules, progressively increasing its chain length. At each alternate step in the chain formation, there is the further possibility of cyclization to give rings containing an even number of atoms (3.18) (six is the smallest ring isolated so far). Almost any possible product can be isolated as a major component by varying the conditions of the reaction, or the proportions of the reactants. The reaction can be carried out in solvents at moderate temperatures, or *neat* at around 130 °C. Ammonium chloride is usually the ammonia source, though free ammonia can be used. Recently, it has been found that the addition of anhydrous metallic salts (e.g. $CoCl_2$, $SnCl_4$, $AlCl_3$) to the reaction mixture increases the rate of formation of phosphazenes. By variations of this type, linear chains containing more than twenty phosphorus atoms, and cyclic compounds containing up to seventeen phosphorus atoms, have been prepared. Compounds of this large and important group are known as phosphonitrilic halides, and their chemistry will be discussed later in this chapter (p. 120).

$$Cl_3PCl_2 + NH_3 \longrightarrow Cl_3P(Cl)\text{—}N^+H_3\ Cl^- \longrightarrow Cl_3P\text{=}NH + 2HCl \quad (3.17)$$

$$Cl_3P\text{=}NH \xrightarrow{Cl_3PCl_2} Cl_3P\text{=}N\text{—}P^+Cl_3\ Cl^-$$

$$Cl_3P\text{=}N\text{—}P^+Cl_3\ Cl^- \xrightarrow{NH_3} Cl_3P\text{=}N\text{—}P^+(Cl)_2\text{—}NH_2\ Cl^-$$

$$Cl_3P\text{=}N\text{—}P^+(Cl)_2\text{—}NH_2\ Cl^- \xrightarrow{-HCl} Cl_3P\text{=}N\text{—}P(Cl)_2\text{=}NH$$

$$Cl_3P\text{=}N\text{—}P(Cl)_2\text{=}NH \xrightarrow{Cl_3PCl_2} Cl_3P\text{=}N\text{—}P(Cl)_2\text{=}N\text{—}P^+Cl_3\ Cl^-$$

$$\xrightarrow{NH_3} \text{cyclic } (N\text{=}PCl_2)(N)(PCl_3)(NH)(PCl_2) \xrightarrow{\text{cyclization}} (NPCl_2)_3 \quad (3.18)$$

$$\xrightarrow[\text{linear route}]{PCl_5} Cl_3P\text{=}N\text{—}PCl_2\text{=}N\text{—}PCl_2\text{=}N\text{—}P^+Cl_3$$

Similar compounds (3.19) are obtained from other phosphorus halides with the limitation of at least three halogen atoms on the phosphorus atom. When

$$nR_2PX_3 + 4nNH_3 \longrightarrow (R_2P\text{=}N\text{—})_n + 3nNH_4^+X^- \quad (3.19)$$

the phosphorus atom has only two substituent halogens, telomerization is impossible, and compounds containing a single phosphorus atom are obtained (3.20). Because of the close relationship of these phosphine imides to phosphonium ylids, their chemistry will be discussed in Chapter 4 (p. 162).

$$R_3PX_2 + R'\text{—}NH_2 \xrightarrow{2Et_3N} R_3P\text{=}NR' + 2[Et_3NH]^+X^- \quad (3.20)$$

phosphine imide

3.1.2 *Phosphorus(V) compounds containing phosphorus–oxygen bonds*

These are compounds of general formula $R_nP(OR')_{5-n}$. Whether they are best represented by a covalent or an ionic structure again depends on the nature of the groups R and R′. Nuclear magnetic resonance, utilizing differences in phosphorus chemical shifts (see p. 43), has been used to some effect in distinguishing between covalent and ionic structures in this instance.

When the ionic structure is dominant (e.g. when $n = 1$ and R is a halogen atom), the compounds usually undergo fairly rapid decomposition by the Arbusov reaction (see p. 54). However, when the structure is covalent (e.g. when $n = 0$), the compounds, although reactive, are often stable enough to be isolated.

In many examples, the pentacovalent phosphorus atom occurs in a five-membered ring (3.21). This type of compound is readily prepared from α-diketones and phosphites (see p. 79) and, according to phosphorus-31 nuclear magnetic resonance evidence, appears to be best represented by the covalent form (3.21). However, there are isolated examples to which the dipolar form (3.22) contributes considerably.

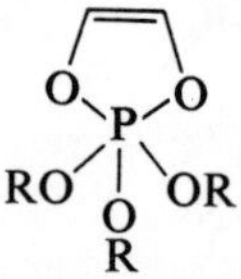

(3.21) 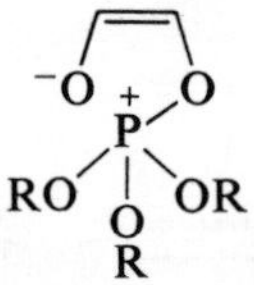 (3.22)

These compounds will themselves react with the α-diketones used in their preparation, although this is slow enough to allow isolation of the first-formed product.

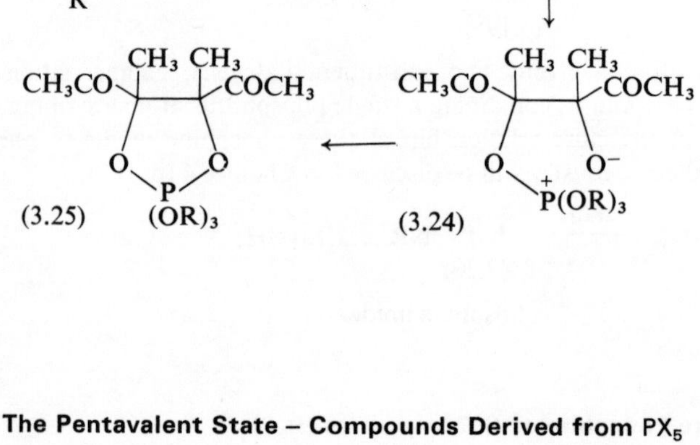

(3.23)

(3.25) (3.24)

The product from this reaction contains the phosphorus atom in a saturated five-membered ring, and its mode of formation is thought to be via (3.23) and (3.24).

Compounds of the type (3.25) are particularly interesting stereochemically as they are able to exist in both a *meso* (3.26) and a *racemic* form (3.27a and b). The five-membered ring is represented as spanning an apical and an equatorial position, since angular strain would be at a minimum under these circumstances (see p. 111). The two forms (3.26) and (3.27) have been separated in a number of cases.

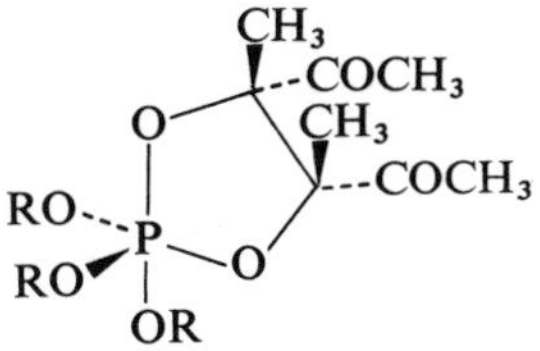

(3.26)

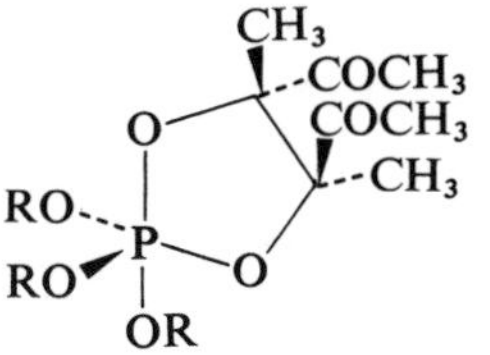

(3.27a)

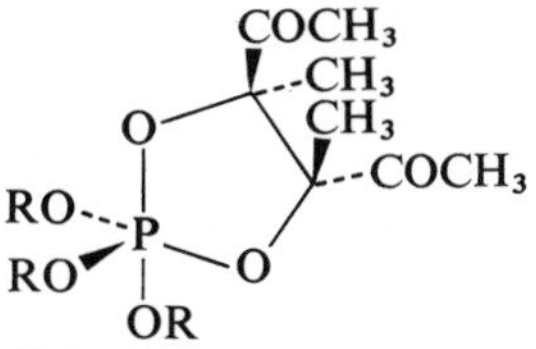

(3.27b)

All these compounds, which are generally solids with fairly low melting points, or liquids, react rapidly with water. This reaction can proceed with either ring opening or ring retention, depending on the structure of the original ring system. In the case of the unsaturated ring system (3.28), the initial nucleophilic substitution at the phosphorus atom is followed by an elimination to form a P═O bond with retention of the ring system.

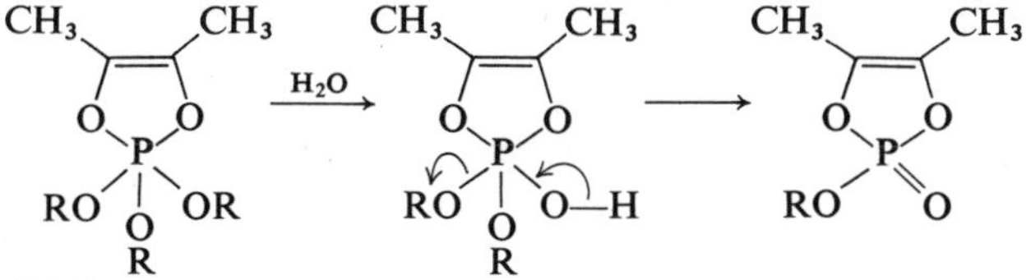

(3.28)

However in the case of the ring systems (3.29) and (3.30), the same initial substitution is followed by ring opening (see p. 112).

$$(RO)_3P(O_2C_2R_2) \xrightarrow{H_2O} (RO)_2P(OH)(O_2C_2R_2) \longrightarrow (RO)_2P(=O)OCHR\text{-}CHR\text{-}O^- \longrightarrow (RO)_2P(=O)CHR\text{-}O^- + RCHO$$

(3.29)

$$(RO)_3P[OC(CH_3)=C(COCH_3)CH(Ph)] \xrightarrow{H_2O} (RO)_2P(OH)[OC(CH_3)=C(COCH_3)CH(Ph)] \longrightarrow (RO)_2P(=O)CH(Ph)C(COCH_3)=C(CH_3)O^- \xrightarrow{H^+} (RO)_2P(=O)CH(Ph)CH(COCH_3)COCH_3$$

(3.30)

The thermal decomposition of these compounds has also been studied. For example, the pentoxyphosphorane (3.31) undergoes rearrangement at 100 °C to give the phosphate (3.32). However, a large number of questions remain to be answered about the reactivity of these pentacovalent-phosphorus systems.

$$CH_3C(O)=C(OEt)O\text{-}P(OR)_3 \xrightarrow{100\,^\circ C} CH_3C(R)(O\text{-}P(O)(OR)_2)\text{-}C(=O)OEt$$

(3.31) (3.32)

3.1.3 *Phosphorus(V) compounds containing five phosphorus–carbon bonds*

These compounds are less stable than other pentavalent-phosphorus compounds and this, together with the difficulties involved in their preparation, has restricted the study of them.

Some members of the series, however, do show reasonable stability. For example pentaphenylphosphorane, Ph_5P, is stable in air at 20 °C and is thermally stable up to 124 °C in nitrogen; it has even been the subject of a patent for flameproofing polystyrene. Since it is also easily prepared, from tetraphenylphosphonium salts (3.33) and phenyllithium, it has been the subject of both structural and chemical studies. X-ray crystallographic studies have confirmed a trigonal-bipyramidal structure (3.34) with the equatorial phenyl groups turned

$$Ph_4P^+X^- + LiPh \longrightarrow Ph_5P + LiX$$

(3.33)

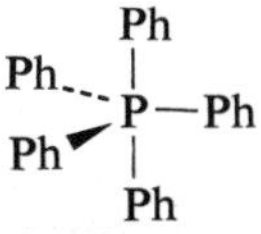

Ph
Ph---P—Ph
Ph
Ph
(3.34)

somewhat out of the basal plane for steric reasons. Pentaphenylphosphorane will react with cold acid to give a tetraphenylphosphonium salt and benzene, presumably by electrophilic substitution at a phenyl nucleus.

$$Ph_5P + H^+ \longrightarrow Ph_4P^+ + PhH$$

Recently, a number of compounds based on the biphenylene nucleus have been prepared by the action of lithium alkyls on the phosphonium salt (3.35). Treatment of the salt (3.35) with $LiAlH_4$ or $NaBH_4$ leads to the phosphorane (3.36; R = H) with four carbon atoms and one hydrogen atom attached to the phosphorus atom. This compound tends to be unstable and decomposes over a

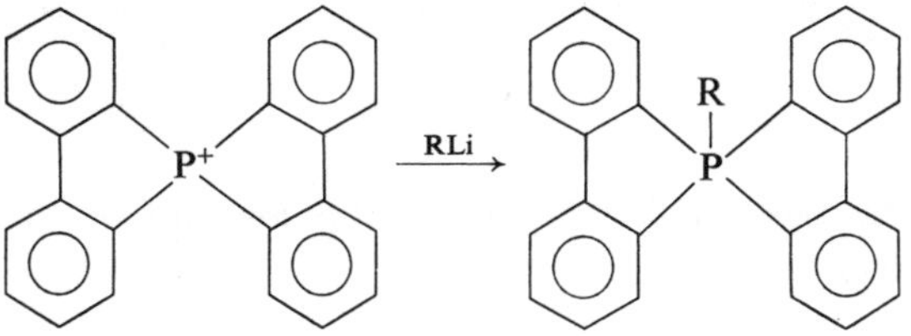

P+ —RLi→ P—R

(3.35) (3.36)

P·

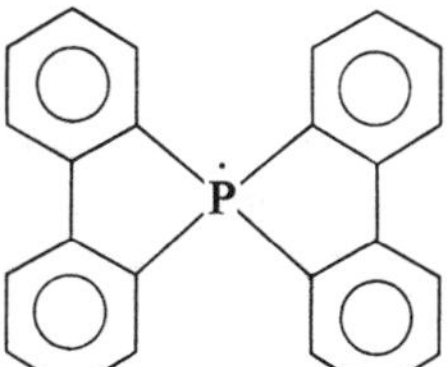

(3.37)

period of weeks at room temperature to give ring-opened products, probably via the radical (3.37). The relative stability of compounds like (3.36; R = H) is probably related to the five-membered spiro ring system (see pp.110–13).

The phosphoranes (3.36) undergo ligand exchange to give (3.38). Support

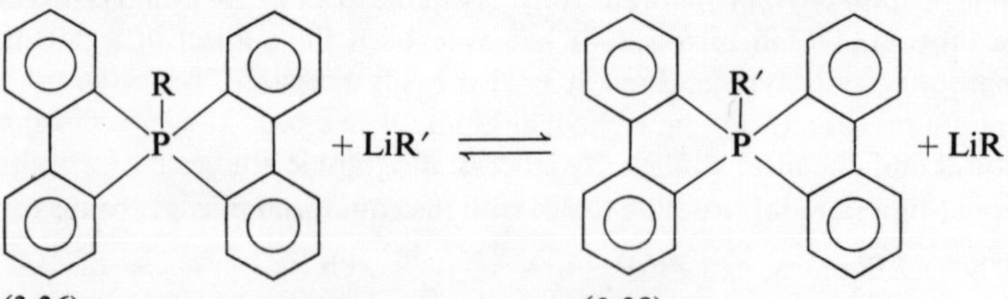

$$(3.36) + LiR' \rightleftharpoons (3.38) + LiR$$

(3.36) (3.38)

for the suggestion that this reaction proceeds via a six-coordinate phosphorus intermediate has been provided by the preparation of stable salts of the type (3.39). These salts can be resolved, which is in full agreement with their postulated octahedral structure (phosphorus pentachloride in the solid state contains a PCl_6^- ion).

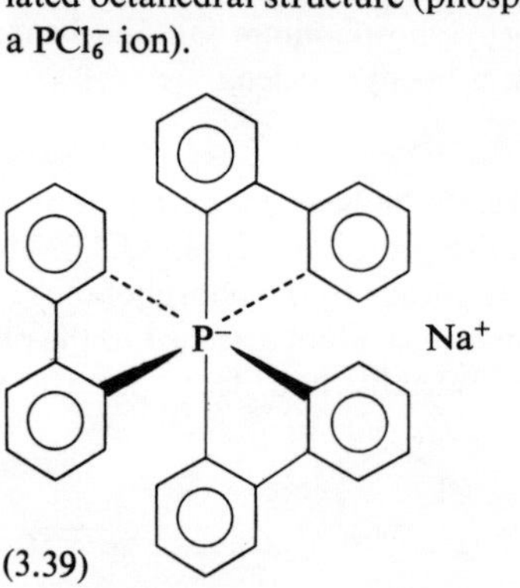

Na^+

(3.39)

There appears to be considerable scope for further investigation of this interesting group of compounds.

3.2 Phosphorus (V) compounds containing the phosphoryl group (P=O)

As might be expected from a consideration of its high bond energy (p. 33), the P═O bond occurs commonly in phosphorus compounds, and in the following pages we shall look at the reactions of compounds containing this functional group. In most cases, analogous compounds containing the P═S bond have been prepared, but these will only be discussed when their chemistry differs sufficiently from the oxygen analogues to merit separate discussion.

3.2.1 *Phosphorus oxy-acids*

The acids are strong, with first ionization constants in the region of 0·5–2·0 (see Table 20), and are often encountered as reagents, for example in dehydration (polyphosphoric acid).

Table 20 Thermodynamic ionization constants of phosphorus oxy-acids*

Acid	*Formula*	pK_1	pK_2
phosphinic acid	$H-P(=O)(H)-OH$	1·1	
phosphorous acid	$H-P(=O)(OH)-OH$	1·3	6·7
monoalkyl phosphite	$RO-P(=O)(H)-OH$	0·8 (R = Et)	
alkanephosphonic acid	$R-P(=O)(OH)-OH$	2·3 (R = Me)	7·9 (R = Me)
phosphoric acid	$HO-P(=O)(OH)-OH$	2·1	7·1
monoalkyl phosphate	$RO-P(=O)(OH)-OH$	(1·8) (R = Et)	7·0 (R = Et)
dialkyl phosphate	$RO-P(=O)(OR)-OH$	(1·5) (R = Et)	

Table 20 – *continued*

Acid	*Formula*	pK_1	pK_2
hypophosphoric acid	$HO-P(=O)(OH)-P(=O)(OH)-OH$	(2·0)	(2·6)
pyrophosphoric acid	$HO-P(=O)(OH)-O-P(=O)(OH)-OH$	1·0	2·0

* Van Wazer (1958), vol. 1, p. 360.
Values in parentheses were crudely estimated from apparent dissociation constants uncorrected for activities.

Their reactivity is in many ways analogous to that of the carboxylic acids. Esterification, of (3.40) for example, takes place via a trigonal-bipyramidal intermediate (3.41), which is formed by attack at a phosphorus atom analogous to attack at the carbonyl carbon atom in carboxylic acids. It is also possible to form acid chlorides (3.42) by reaction with phosphorus pentachloride.

$$R_2P(=O)-OH + R'OH \longrightarrow R_2P(O^-)(O^+H_2)(OR') \longrightarrow R_2P(=O)-O-R$$

(3.40) (3.41) alkyl dialkylphosphinate + H_2O

The anions of phosphorus oxy-acids show nucleophilic reactivity, a property which is shown by the acids themselves under favourable circumstances. For example, the oxy-acid (3.43) reacts with acetylenic ethers to give the vinyl ester (3.44), presumably by the route shown.

$$R_2P(=O)OH + PCl_5 \xrightarrow{-HCl} R_2P(=O)-O-PCl_4 \;\; Cl^- \longrightarrow R_2P(=O)Cl$$

(3.42) dialkylphosphin chloride + $Cl_3P{=}O$

$$(RO)_2P(=O)\text{—}O\text{—}H + H\text{—}C\equiv C\text{—}\ddot{O}\text{—}R' \longrightarrow (RO)_2P(=O)\text{—}O^- + H_2C{=}C{=}\dot{O}^+\text{—}R'$$

(3.43)

$$\downarrow$$

$$(RO)_2P(=O)\text{—}O\text{—}C(=CH_2)\text{—}O\text{—}R'$$

(3.44)

At first sight, the simplest phosphorus acids (3.45 and 3.46) would appear to contain trivalent phosphorus and, until recently, there was considerable controversy as to their real structure. It is now evident however that the alternate

$R_2P\text{—}OH$ (3.45) $R\text{—}P(OH)_2$ (3.46)

tautomeric forms (3.47 and 3.48) predominate (>99%), again presumably because of the P=O bond energy. These compounds still show acidic properties, although in the case of (3.47) the pK_a value is much higher than those of the

$$R_2P\text{—}O\text{—}H \rightleftharpoons R_2P(=O)H$$

(3.45) (3.47) dialkylphosphinous acid (dialkylphosphine oxide)

$$RP(O\text{—}H)_2 \rightleftharpoons RP(=O)(OH)H$$

(3.46) (3.48) alkylphosphonous acid

oxy-acids. This acidity is due to the electron-withdrawing effect of the phosphoryl bond (compare with the aldehydes). The compounds will form salts (3.49) with sodium alkoxides, and add to activated double and triple bonds to give tertiary phosphine oxides (3.50).

$$R_2P(=O)\text{—}H + R'ONa \longrightarrow R_2P(=O)^-Na^+ + R'OH$$

(3.49)

$$R_2P(=O)H + CH_2{=}CH\text{—}CN \xrightarrow[\text{reflux}]{\text{benzene}} R_2P(=O)\text{—}CH_2\text{—}CH_2\text{—}CN \quad 70\text{–}90\%$$

(3.50)

Compelling evidence, mainly from kinetics, has suggested that the tautomer (3.45) containing trivalent phosphorus is a reaction intermediate in several cases. At room temperature, however, its contribution to the above equilibrium must be very small, since optically active benzylphenylphosphine oxide (3.51) is configurationally stable in methanol at this temperature. It would be interesting to see if the anion derived from (3.51) has any configurational stability. This seems possible as deuterium exchange takes place in methanol-D (CH_3OD) with retention of the configuration at the phosphorus atom.

$$\text{Ph—CH}_2\text{—P(=O)(H)(Ph)}$$

(3.51)

$$\text{Ph—CH}_2\text{—P(=O)(H)(Ph)} + CH_3OD \rightleftharpoons \text{Ph—CH}_2\text{—P(=O)(D)(Ph)} + CH_3OH$$

Phosphorus oxy-acids and their derivatives play a vital role in biological systems (see Chapter 5), but until recently no compound containing a phosphorus–carbon bond had been isolated from a living organism. It is perhaps not surprising that the first of these compounds to be isolated was a phosphorus oxy-acid. The biosynthetic route to this compound, 2-aminoethylphosphoric acid (3.52), has not yet been fully resolved. However it is now thought that the carbon–phosphorus bond is initially formed via a cyclic rearrangement of the enol ester (3.53), followed by enzymic amination and decarboxylation.

$$(HO)_2P(=O)\text{—CH}_2\text{—CH}_2NH_2$$

(3.52)

A summary of the phosphorus oxy-acids and the routes to them via the hydrolysis of phosphorus chlorides can be found in Table 21.

$$\underset{(3.53)}{\text{H—O(R—O)P(=O)—O—C(=CH}_2\text{)COOH}} \longrightarrow \text{H—O(R—O)P(=O)—CH}_2\text{—C(=O)—COOH}$$

$$\downarrow$$

$$\text{HO(RO)P(=O)—CH}_2\text{—CH}_2\text{—NH}_2 \xleftarrow{-CO_2} \text{H—O(R—O)P(=O)—CH}_2\text{—CH(NH}_2\text{)—COOH}$$

Table 21 Reactions of phosphorus halides

Phosphorus halide	*With water*	*With alcohols* R′OH	*With amines* R'_2NH
R_2PCl dialkylphosphinous chloride	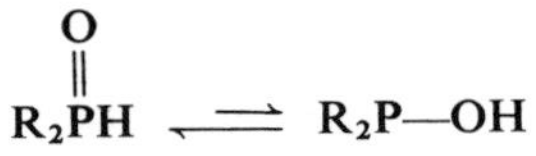 $R_2P(=O)H \rightleftharpoons R_2P{-}OH$ dialkylphosphinous acid (dialkylphosphine oxide)	$R_2P{-}OR'$ alkyl dialkylphosphinate	$R_2P{-}NR'_2$ tetraalkylphosphinous amide
$RPCl_2$ alkylphosphonous dichloride	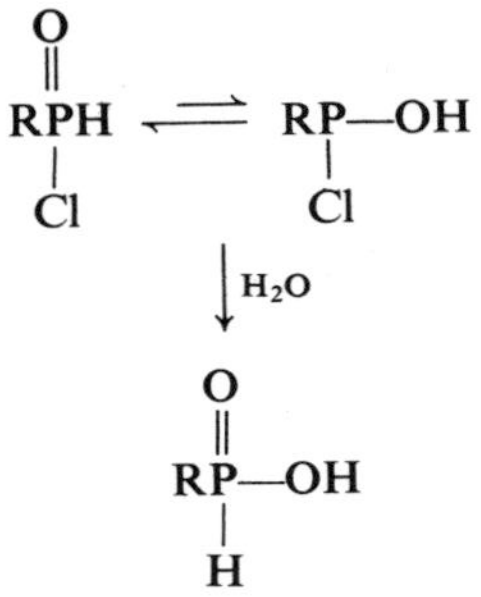 $RP(=O)H(Cl) \rightleftharpoons RP(Cl){-}OH$ $\downarrow H_2O$ $RP(=O)(H){-}OH$ alkylphosphinic acid	$RP(OR')_2$ dialkyl alkylphosphonite	$RP(NR'_2)_2$ *N*,*N*,*N*,*N*-tetraalkyl-P-alkylphosphonous diamide

Table 21 – *continued*

Phosphorus halide	*With water*	*With alcohols* R′OH	*With amines*
PCl_3 phosphorous chloride	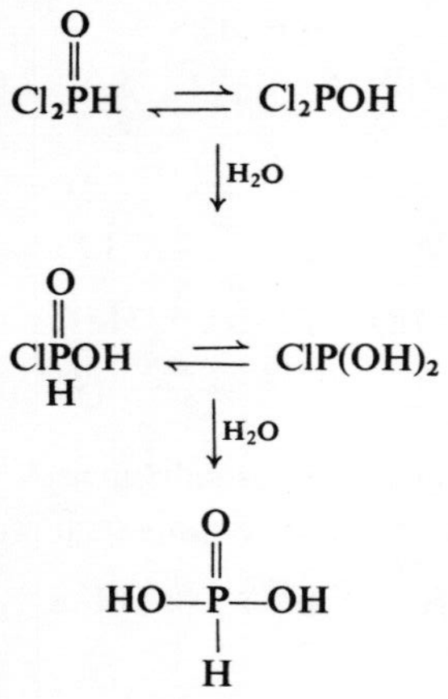 $Cl_2P(=O)H \rightleftharpoons Cl_2POH$ $\downarrow H_2O$ $ClP(=O)(H)OH \rightleftharpoons ClP(OH)_2$ $\downarrow H_2O$ $HO{-}P(=O)(H){-}OH$ phosphonic acid	$(R'O)_3P$ trialkyl phosphite	$(R'_2N)_3P$ hexaalkyl phosphorous triamide
$R_4P^+Cl^-$ tetraalkyl-phosphonium chloride	$R_4P^+OH^-$ $\downarrow$ $R_3P{=}O$ trialkylphosphine oxide	$R_4P^+OR^-$ phosphonium alkoxide	

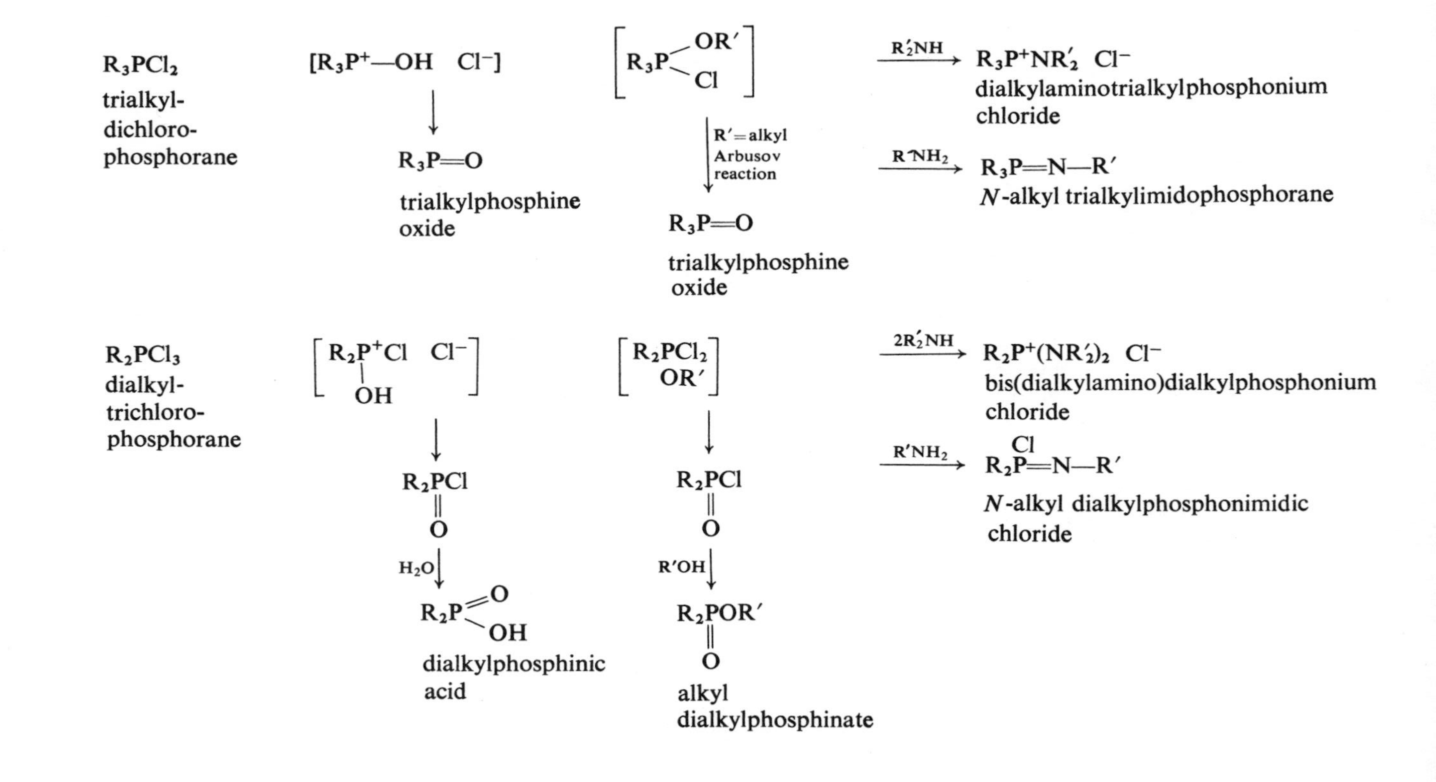
R₃PCl₂
trialkyl-
dichloro-
phosphorane
[R₃P⁺—OH Cl⁻]
R₃P=O
trialkylphosphine
oxide
R₃P(OR′)Cl
R′=alkyl
Arbusov
reaction
R₃P=O
trialkylphosphine
oxide
R′₂NH
R₃P⁺NR′₂ Cl⁻
dialkylaminotrialkylphosphonium
chloride
R′NH₂
R₃P=N—R′
N-alkyl trialkylimidophosphorane
R₂PCl₃
dialkyl-
trichloro-
phosphorane
R₂P⁺Cl(OH) Cl⁻
R₂PCl=O
H₂O
R₂P(=O)OH
dialkylphosphinic
acid
R₂PCl₂OR′
R₂PCl=O
R′OH
R₂POR′=O
alkyl
dialkylphosphinate
2R′₂NH
R₂P⁺(NR′₂)₂ Cl⁻
bis(dialkylamino)dialkylphosphonium
chloride
R′NH₂
R₂P(Cl)=N—R′
N-alkyl dialkylphosphonimidic
chloride

Table 21 – *continued*

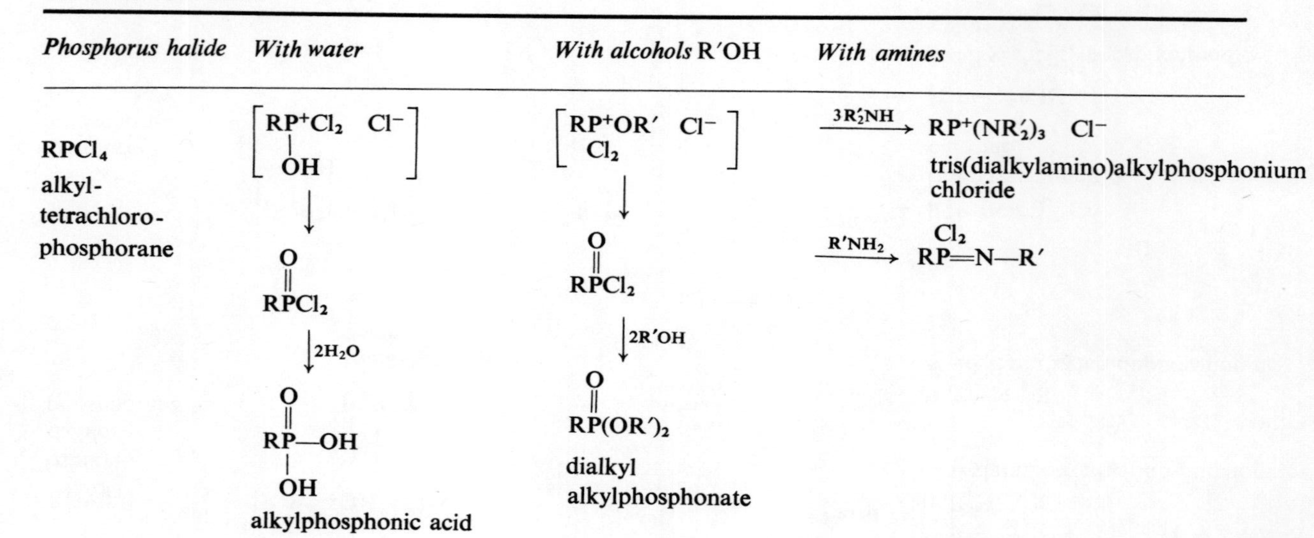

Phosphorus halide	*With water*	*With alcohols* R′OH	*With amines*
$RPCl_4$ alkyl-tetrachloro-phosphorane	$[RP^{+}Cl_2(OH)\ Cl^{-}]$ $\downarrow$ $RP(=O)Cl_2$ $\xrightarrow{2H_2O}$ $RP(=O)(OH)-OH$ alkylphosphonic acid	$[RP^{+}OR'Cl_2\ Cl^{-}]$ $\downarrow$ $RP(=O)Cl_2$ $\xrightarrow{2R'OH}$ $RP(=O)(OR')_2$ dialkyl alkylphosphonate	$\xrightarrow{3R'_2NH}$ $RP^{+}(NR'_2)_3\ Cl^{-}$ tris(dialkylamino)alkylphosphonium chloride; $\xrightarrow{R'NH_2}$ $RPCl_2{=}N{-}R'$

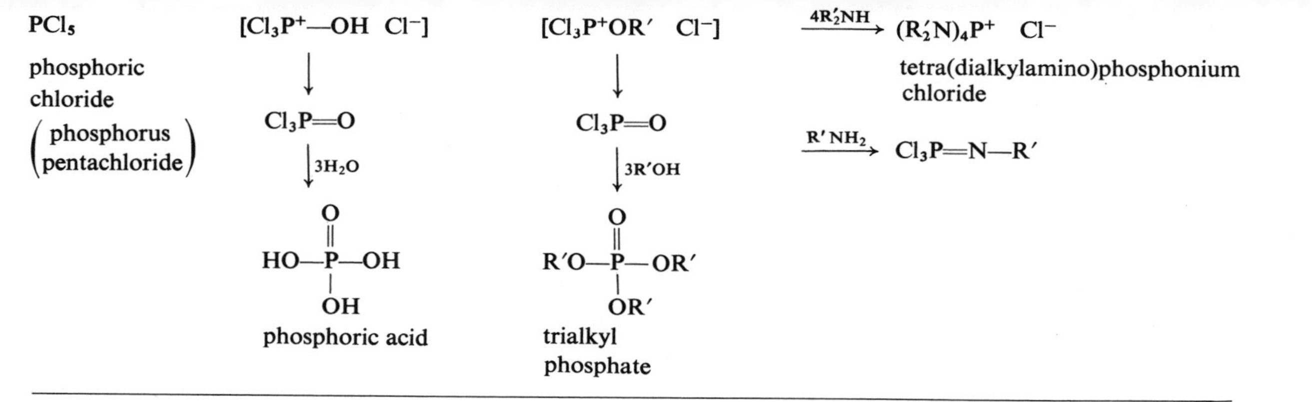

Phosphorus halide

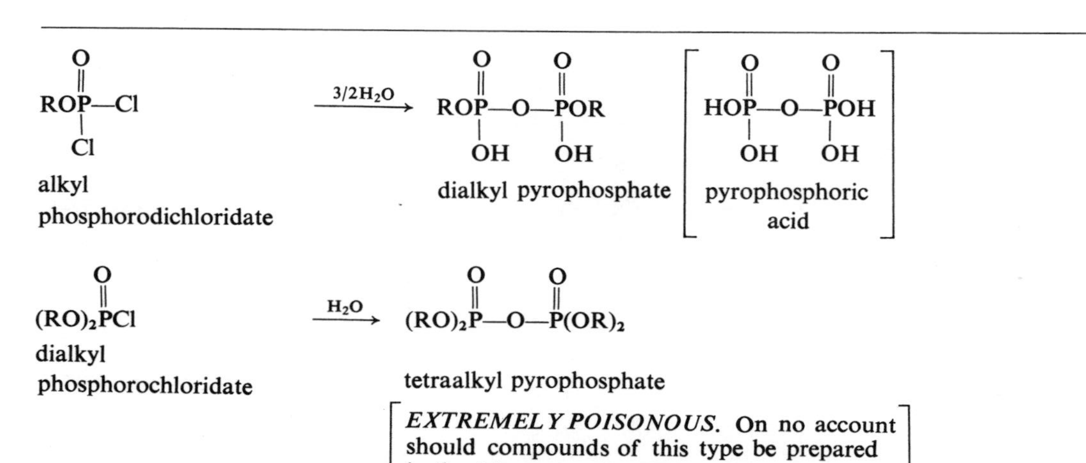

EXTREMELY POISONOUS. On no account should compounds of this type be prepared in the laboratory. See Chapter 5, pp. 200-206.

Table 21 – *continued*

Phosphorus halide		
$ROP(=O)Cl_2$	$\xrightarrow{HOP(=O)(OR')_2}$	$(R'O)_2P(=O)-O-P(=O)(OR)-O-P(=O)(OR')_2$ pentaalkyl triphosphate
$Cl-P(=O)(OR)-Cl$	$\xrightarrow{HO-P(=O)(OR)-OH}$	$\left[-P(=O)(OR)-O-P(=O)(OR)-O-\right]_n$ polyphosphoric acid (R=H)
$Cl-P(=O)(Cl)-Cl$	$\xrightarrow{3(RO)_2P(=O)OH}$	$RO-P(=O)(OR)-O-P(=O)(O-P(=O)(OR)-O-R)-O-P(=O)(OR)-O-R$ hexaalkyl tetraphosphate

3.2.2 *Derivatives of phosphorus oxy-acids*

The phosphorus oxy-acids form the expected derivatives: acid halides, esters, amides, anhydrides, etc., and their thioanalogues. Examples of these are included in Table 21.

Phosphine oxides $R_3P{=}O$ have been included in this table, although they are not strictly derivatives of phosphorus oxy-acids. However, their chemistry is closely allied to that of the oxy-acids owing to the overriding influence of the phosphonyl group.

The reactions of all these derivatives can be separated into two types: those which involve reaction at the phosphorus atom, and those which do not.

(a) *Reactions involving a central phosphorus atom.* In most examples of this type of reaction, the phosphorus atom acts as an electrophile. Reactions which appear to be due to nucleophilic phosphorus invariably involve a potentially trivalent phosphorus atom (see p. 48).

There are many examples of this type of reaction, but they can all be represented, at least formally, by a nucleophilic displacement at the phosphorus atom.

$$A^- + \underset{(3.55)}{\overset{O}{\overset{\|}{\underset{/\,|\,\backslash X}{P}}}} \longrightarrow \overset{O}{\overset{\|}{\underset{/\,|\,\backslash A}{P}}} + X^-$$

(3.54) (3.55)

The phosphoryl compounds (3.55) which have been the object of most studies are the esters ($X = OR$), the halides ($X =$ halide) and the amides ($X = NR_2$). The nucleophiles (3.54) involved have commonly been alcohols, water and amines, although almost any nucleophile will carry out the reaction. These reactions are extremely important because living organisms rely on them for both energy conversion and protein synthesis (see Chapter 5).

Our generalized reaction between (3.54) and (3.55) has been widely used in this context to prepare phosphate esters (known as *phosphorylation*). Thus many workers from different fields have provided information about these reactions, and we know more about these mechanisms than any others in phosphorus chemistry.

Before we look at some of the more interesting aspects of the chemistry of the phosphoryl group, we will consider its more general points.

The effectiveness of any nucleophile depends on a number of factors, but primarily on (a) its basicity (b) its polarizability and (c) any conjugative effects within the nucleophile (the so-called α-effect). The phosphoryl group is similar to the carbonyl group in that the ease of attack depends almost entirely on the basicity of the nucleophile.

In the case of the leaving group, the two functions differ. The nature of the leaving group is much more important in attack at a phosphoryl group than at a carbonyl group. This is presumably because nucleophilic attack at a carbonyl group involves a stable tetrahedral intermediate (3.58), whereas attack at a phosphoryl group involves a less stable species (3.59) (but see p. 111), and is probably trigonal bipyramidal in structure.

$$R'\text{—}C(=O)\text{—}O\text{—}R \; (3.56) \; + \; OH^- \rightleftharpoons R\text{—}C(O^-)(OR)(OH) \; (3.58)$$

$$(RO)_3P{=}O \; (3.57) \; + \; OH^- \rightleftharpoons O^-\text{—}P(OR)_3(OH) \; (3.59)$$

The carbonyl group and the phosphoryl group themselves differ in structure; while carbonyl compounds are planar (3.56), phosphoryl compounds are tetrahedral (3.57). This difference in structure provides a possible explanation of the entropy change on hydrolysis, which is much more negative for phosphoryl esters than for carboxylic esters.

Since symmetry considerations make delocalization via a phosphorus atom impossible (see p. 40), substituents on the phosphorus atom generally have much less effect on reaction than do those at a carbonyl centre. The effect on reaction rates can differ by as much as 10^3 between the phosphonyl and carbonyl groups.

This difference is perhaps better understood in terms of canonical forms. One contributing structure of the carbonyl group is (3.60), but the corresponding contributor to the phosphoryl group (3.61) cannot exist, since the symmetry of the 3d orbitals on the phosphorus atom precludes overlap with both oxygen orbitals at the same time.

$$R\text{—}C(=O)\text{—}O\text{—}R \longleftrightarrow R\text{—}C(O^-){=}O^+\text{—}R' \; (3.60)$$

$$RO\text{—}P(=O)(OR)_2 \; \longleftrightarrow\!\!\!\!\!\!\times\;\; R\text{—}O^+{=}P(O^-)(OR)_2 \; (3.61)$$

Although the reactions that we have been considering all involve a replacement of one group by another at the phosphorus atom, there are a number of different pathways by which this displacement can take place.

(i) Addition–elimination. Firstly, the nucleophile can attack the phosphoryl phosphorus atom to give a phosphorus(V) intermediate (3.62), which then

$$R_2P(=O)-X + Y^- \longrightarrow R_2P(O^-)(X)(Y) \longrightarrow R_2P(=O)Y + X^-$$

(3.62) (3.63)

proceeds to eliminate the leaving group to give the product (3.63). The reaction takes place in two discrete steps and is said to involve an addition–elimination mechanism. This mechanism is the exact analogue of reactions at a carbonyl group and, as we have seen (p. 108), the intermediate (3.62) tends to be less stable than the equivalent intermediate derived from attack at a carbonyl group. It is therefore not surprising that the addition–elimination mechanism only occurs when the intermediate (3.62) is especially stable (e.g. when it contains phosphorus in a five-membered ring, p. 111), or when the leaving group is particularly poor (in which case the reaction may not take place at all).

Labelling techniques have been used effectively to study this mechanism. If a base-catalysed hydrolysis reaction is carried out in $H_2{}^{18}O$ and the above mechanism operates, an intermediate (3.64) with an appreciable lifetime will be formed. If this is in equilibrium with the starting material (3.65), an exchange process will take place with replacement of the phosphoryl oxygen atom by oxygen-18. Hydrolysis of this ^{18}(3.65) by $H_2{}^{18}O$ will then lead to a product containing more than one oxygen-18 atom. Studies of this type have been used to exclude the addition–elimination pathway from a number of hydrolysis reactions, since the products have invariably contained only one oxygen-18 atom.

Optically active phosphoryl compounds have also been used in a study of these mechanisms and, in all cases, reaction has been accompanied by an inversion of

$$(RO)_3P{=}O \ (3.65) \xrightarrow{^{18}OH^-} (RO)_3P(O^-)(^{18}OH)\ (3.64) \xrightarrow{-OH^-} (RO)_3P{=}{}^{18}O\ {}^{18}(3.65)$$

$$(3.64) \xrightarrow{-O\text{-}R} (RO)_2P(=O)(^{18}OH)$$

$$^{18}(3.65) \xrightarrow{^{18}OH^-} (RO)_2P(^{18}O^-)(OR)(^{18}OH) \xrightarrow{-OR^-} (RO)_2P(={}^{18}O)(^{18}OH)$$

the configuration at the phosphorus atom (see however p. 133). This is evidence for a trigonal-bipyramidal intermediate (3.66) because, if five-coordinate, only an intermediate with this geometry can lead to an inversion of the configuration. Moreover, even with this geometry, inversion will only take place if the entering and leaving groups are either both apical or both equatorial (see p. 131).

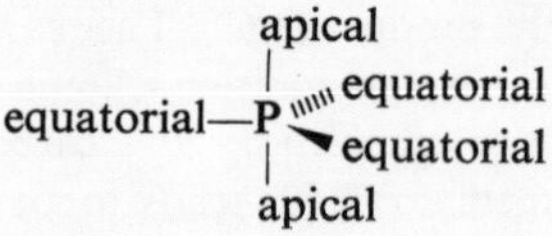

apical — equatorial—P — equatorial — equatorial — apical

(3.66)

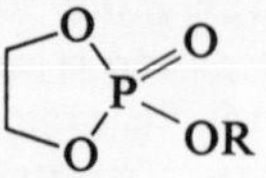

(3.67)

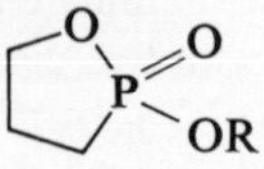

(3.68)

In five-membered cyclic esters like (3.67) and (3.68), only an addition–elimination mechanism appears to provide an explanation of the experimental observations. For both hydrolysis and oxygen exchange, these compounds show enormous increases in reaction rate over their acyclic analogues.

In the case of the diester (3.69) the rates of both hydrolysis and oxygen exchange are 10^8 times greater than for dimethyl phosphate (3.70). This rate increase is not confined to ring-opening reactions. When the cyclic phosphate (3.71) is hydrolysed, the rate is again about 10^8 times greater than with trimethyl phosphate ($(CH_3O)_3P{=}O$), and both the ring-opened (3.72) and ring-retained (3.73) products are produced. The relative yields (70 and 30 per cent) show that both reactions (ring opening and retention) proceed at comparable rates, and so the

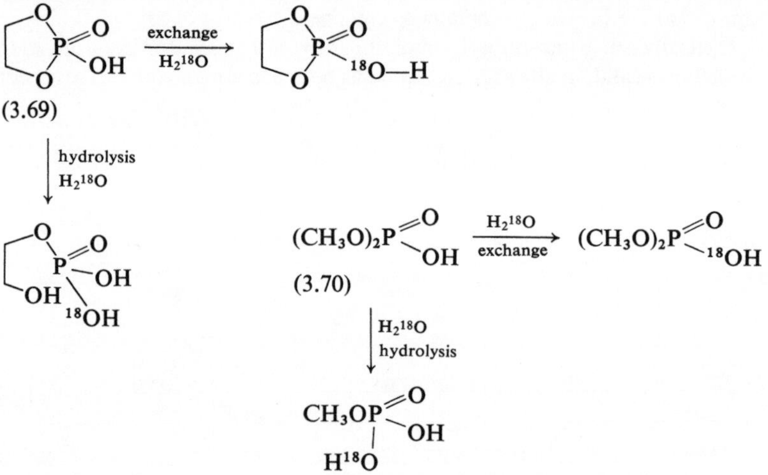

(3.69) —exchange, $H_2{}^{18}O$→ cyclic P(=O)^{18}O–H

(3.69) —hydrolysis, $H_2{}^{18}O$→ ring-opened P(=O)(OH)(^{18}OH) with CH$_2$OH

$(CH_3O)_2P(=O)OH$ (3.70) —$H_2{}^{18}O$, exchange→ $(CH_3O)_2P(=O){}^{18}OH$

(3.70) —$H_2{}^{18}O$, hydrolysis→ $CH_3OP(=O)(OH)(H^{18}O)$

(3.71) —H_2O, pyridine→ ring-opened P(=O)(OH)(OCH_3) 70% 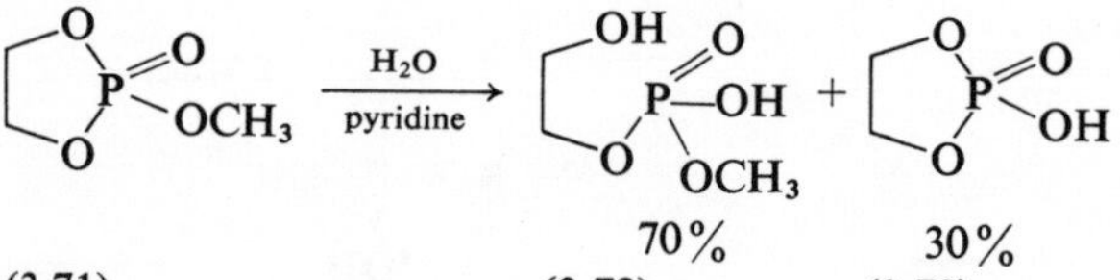 (3.72) + cyclic P(=O)OH 30% (3.73)

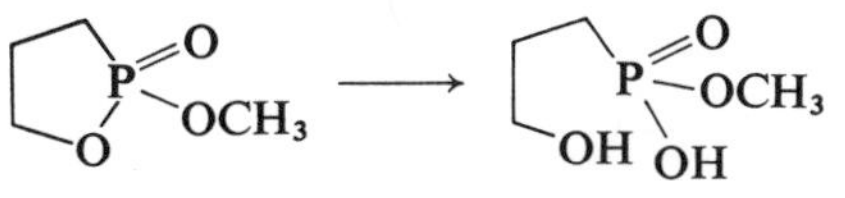

(3.74) $\longrightarrow$ (3.75)

rate increase is not confined to hydrolysis of the cyclic ester link. However, hydrolysis of the ester (3.74), although still accelerated by some 10^6 times, leads only to the ring-opened product (3.75). Any explanation of these enormous rate differences must also account for these results.

By far the most satisfactory solution to this problem comes from work by Westheimer and Haake who suggested that the hydrolysis reactions proceeded via a trigonal-bipyramidal intermediate (the large number of pentavalent-phosphorus compounds isolated possessing a five-membered ring, suggests some stability in this) (see p. 79). From a study of many results obtained from hydrolysis of various cyclic and acyclic phosphorus esters they deduced four hypotheses relating to the intermediate (3.66):

(a) The most electronegative elements will prefer to be apical.

(b) The five-membered ring will be more stable bridging apical–equatorial positions, with an O—P—O bond angle of 90°, rather than bridging equatorial–equatorial positions, with an angle of 120°.

(c) Pseudo-rotation (see p. 31) takes place as long as (d) and (b) are preserved.

(d) Entering groups enter apically, and leaving groups leave apically, even if this requires several pseudo-rotations in between.

Westheimer and Haake then tried to explain the observed phenomena. For hydrolysis of the phosphate ester (3.71), if the entering group is apical (d), and the five-membered ring spans equatorial and apical positions (b), the intermediate should have the stereochemistry (3.76). The formation of this intermediate relieves the strain, because the five-membered-ring bond angle at the phosphorus atom is reduced to 90° from the near-tetrahedral (108°) value in (3.71). This relief causes all the rate accelerations observed. However, the five-membered ring must not span any positions other than equatorial and axial ones in the route from (3.76) to any of the products. If there were equatorial–equatorial bridging at any stage, the strain would not be relieved, and the route to the particular product would show no corresponding acceleration.

This is all achieved by a series of pseudo-rotations. The route to the product showing hydrolysis with ring opening (3.77) is simple, since the intermediate (3.76) has a ring oxygen in an apical position – see (d). The route to the product showing hydrolysis with ring retention (3.79) involves a single pseudo-rotation (c) to give the intermediate (3.78); this retains apical–equatorial ring bridging (b), and can now lose a methoxyl group from an apical position to give the product (3.79). Finally, phosphoryl oxygen exchange can occur: starting from (3.76), a single pseudo-rotation gives (3.80) (with retention of ring bridging); this is followed by protonation and the loss of an apical hydroxyl group to give the oxygen-exchanged product (3.81).

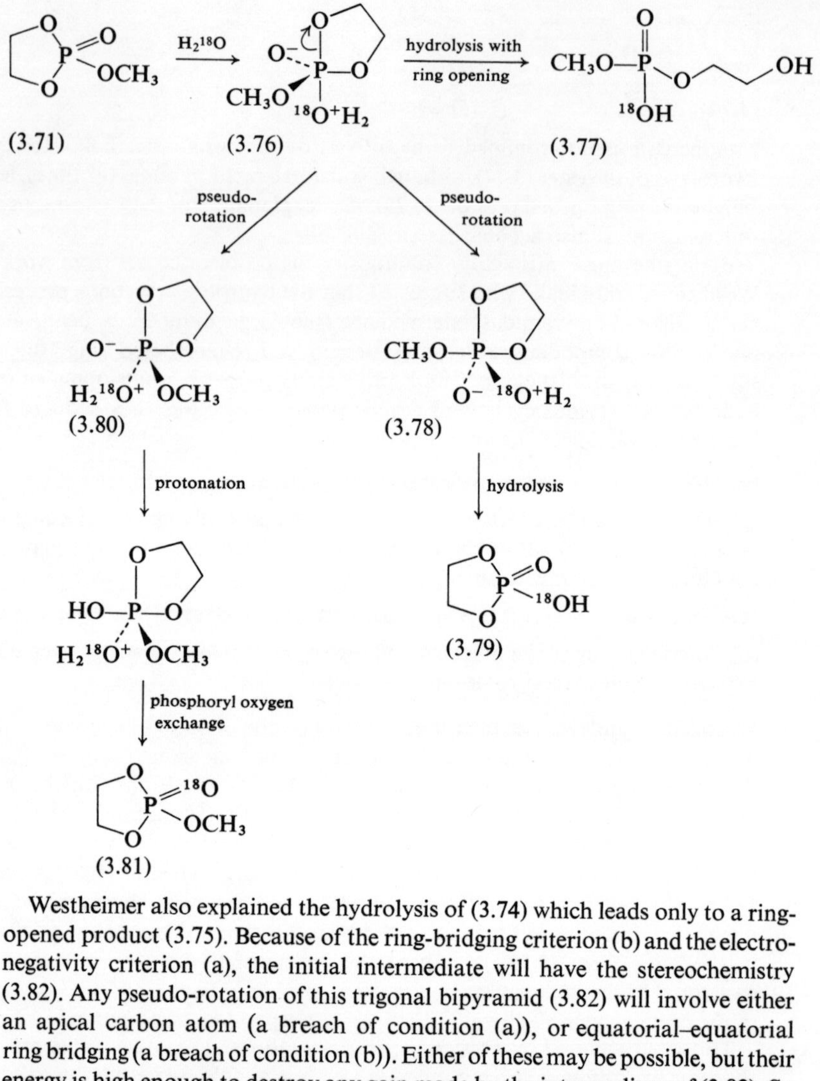

Westheimer also explained the hydrolysis of (3.74) which leads only to a ring-opened product (3.75). Because of the ring-bridging criterion (b) and the electronegativity criterion (a), the initial intermediate will have the stereochemistry (3.82). Any pseudo-rotation of this trigonal bipyramid (3.82) will involve either an apical carbon atom (a breach of condition (a)), or equatorial–equatorial ring bridging (a breach of condition (b)). Either of these may be possible, but their energy is high enough to destroy any gain made by the intermediacy of (3.82). So the formation of (3.75) by apical ring opening in (3.82) will exhibit acceleration, but exocyclic hydrolysis (to give 3.83) will be very much slower.

The great strength of Westheimer's theory lies in its ability to predict rate increases, or the lack of them, in several different types of cyclic compounds. The theory has also been applied to cyclic phosphonium salts (see pp. 131–3).

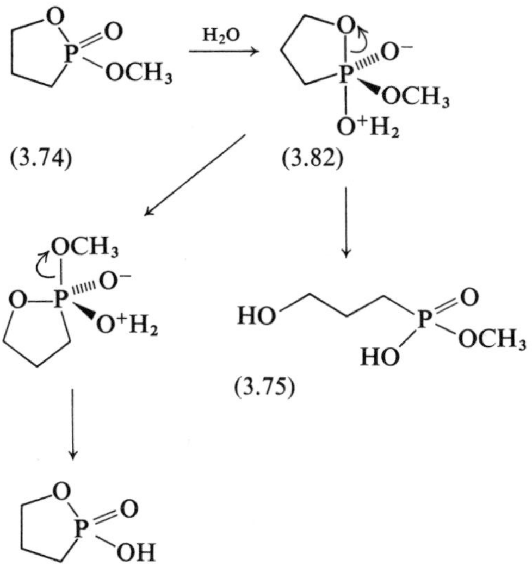

(3.74)

(3.82)

(3.83)

(3.75)

(ii) Elimination–addition. Like the addition–elimination mechanism, this involves the formation of a relatively stable intermediate (3.84), but by elimination rather than addition. The elimination is followed by addition of the nucleophile to give the product (3.85).

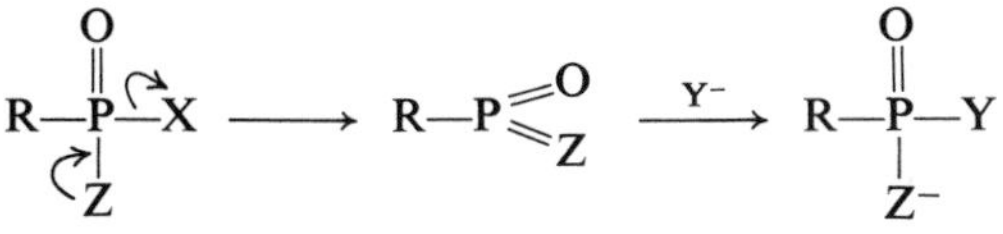

(3.84) (3.85)

X and Z = O, S or NR

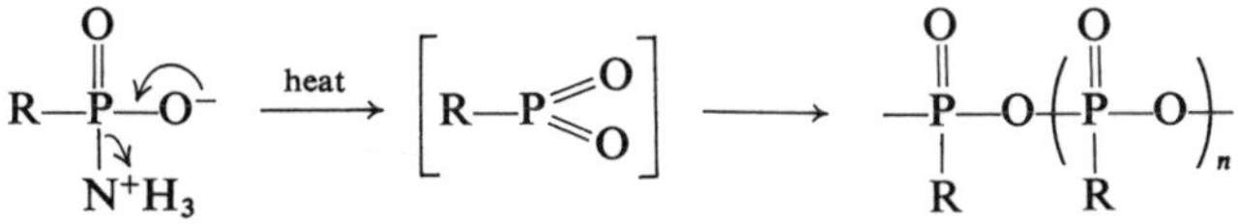

heat

(3.86)

All attempts to isolate compounds of type (3.84) have failed, and the products are usually polymeric (e.g. 3.86). It is interesting to consider this in the light of the extreme reluctance of trivalent-phosphorus compounds to form p_π–p_π double bonds (see p. 19). This of course does not preclude the existence of such compounds as intermediates in reactions, and there is good evidence that such intermediates do exist thus. Elimination–addition is probably quite a common pathway in hydrolysis reactions, especially those carried out in a strong base. Most examples of this type of mechanism involve an anion (e.g. 3.87); this is reasonable in view of the inevitable difficulty of nucleophilic attack at a negatively charged centre which leads to the suppression of other mechanisms.

Lack of discrimination in some reactions between different nucleophiles is the most compelling evidence for the operation of a mechanism of this type. For example, if the phosphate anion (3.87) is hydrolysed in a mixture of water and methanol, the ratio of the products depends entirely on the proportion of each component in the solvent. This suggests that the reaction involves a reactive intermediate, since it takes no account of the differing nucleophilicity of methanol and water. The most obvious intermediate to suggest is (3.88), particularly as this is known to be highly reactive.

$$O_2N\text{-}C_6H_4\text{-}O\text{-}P(=O)(O^-)\text{-}OH \longrightarrow O^-\text{-}PO_2 \;\; \xrightarrow{H_2O} O^-\text{-}P(=O)(OH)\text{-}OH \;;\; \xrightarrow{MeOH} O^-\text{-}P(=O)(OH)\text{-}OMe$$

(3.87) (3.88)

$$(PrNH)_2P(=O)Cl \xrightarrow[K_1]{H_2O} (PrNH)_2P(=O)OH$$

(3.89)

$$(Me_2N)_2P(=O)Cl \xrightarrow[K_2]{H_2O} (Me_2N)_2P(=O)OH$$

(3.90) tetramethylphosphorodiamidic acid

Similar evidence has been obtained from hydrolysis of the phosphoryl halides (3.89 and 3.90) in the presence of various catalysts. When the catalyst is pyridine or fluoride (weak bases), the ratio of the rate constants K_1 and K_2 is about unity, but when the catalyst is hydroxide ion, the ratio of K_1 and K_2 becomes 10^6. This suggests that in the presence of a strong base the two substances are hydrolysed by different paths. An attractive alternative for this mechanism which would apply to (3.89) and not to (3.90), thus explaining the $K_1 : K_2$ ratios, would be an elimination–addition via (3.91).

$$(PrNH)_2P(=O)Cl \xrightarrow{^-OH} PrNH\text{-}P(=O)=N\text{-}Pr \xrightarrow{H_2O} PrNH\text{-}P(=O)(OH)\text{-}NHPr$$

(3.91)

(iii) Direct displacement. Direct displacement is analogous to the S_N2 mechanism at a saturated carbon atom. The attacking nucleophilic group displaces the leaving group via a transition state (3.92), without the formation of an intermediate.

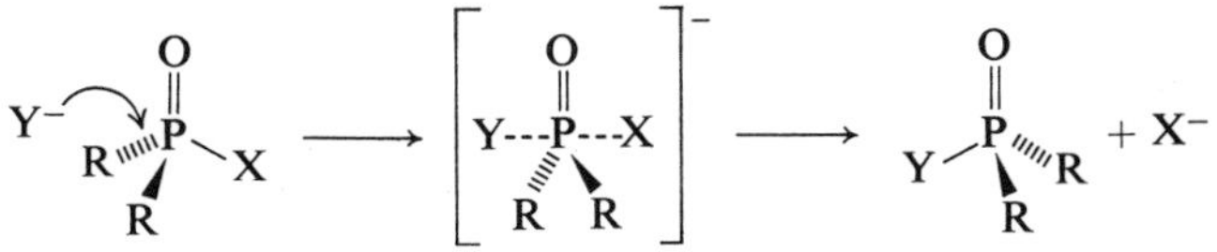

(3.92)

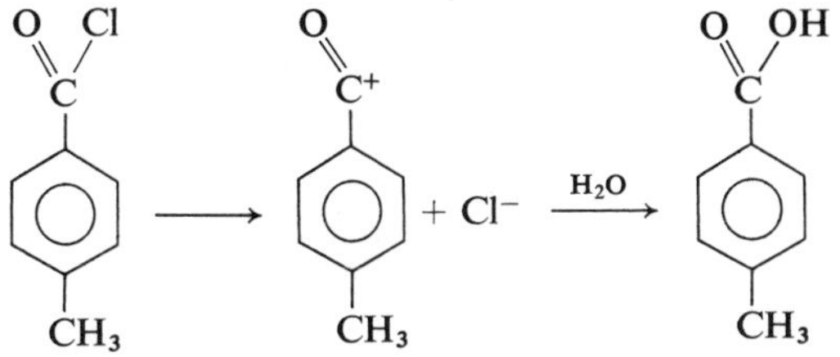

(3.93) (3.94)

Examples of largely unimolecular substitutions at carbonyl groups are known. *p*-Methylbenzoyl chloride (3.93), for example, appears to undergo hydrolysis in aqueous ethanol by a largely S_N1 mechanism, presumably because the positive charge can be spread over the electron-rich aromatic ring in (3.94). The greater polarization of the phosphoryl group ($P^{\delta+}$–$O^{\delta-}$) should reduce the contribution from a unimolecular mechanism which would involve (3.95). This is indeed the case, although there are a few examples of what appear to be unimolecular substitutions. One of these, the hydrolysis of the halide (3.96), illustrates the importance of solvent effects: it appears to change from a bimolecular substitution to a unimolecular one when the solvent is changed from aqueous acetone to formic acid.

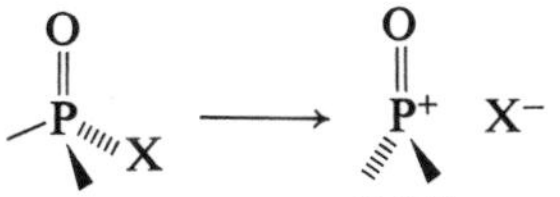

(3.95) 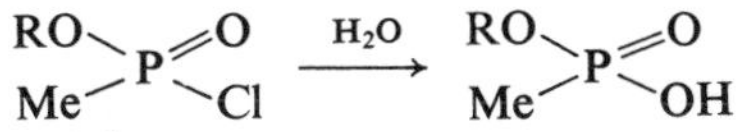 (3.96)

alkyl methylphosphonic acid

A variety of evidence is available for a direct substitution mechanism. Hydrolysis of trimethyl phosphate with $^{18}OH^-$ has shown that both an attack at a phosphorus atom and P—O bond cleavage are involved. Since only one oxygen-18 atom is incorporated in the product, the alternative addition–elimination mechanism discussed above (p. 108) is, in this case, unlikely. Evidence for a bimolecular substitution mechanism comes also from reactions with a mixture of nucleophiles, since reactions following an S_N2 mechanism tend to discriminate on the basis of nucleophilicity.

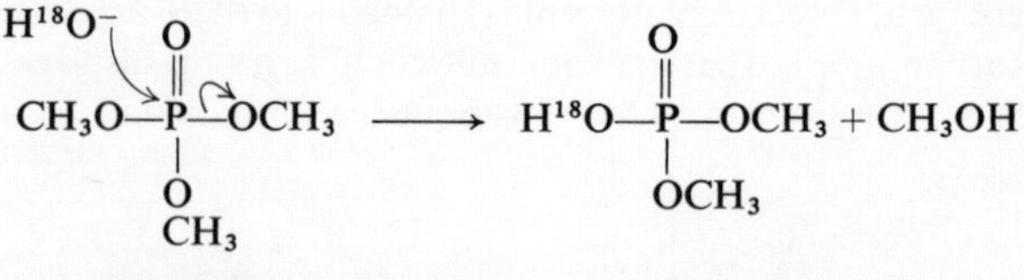

$$H^{18}O^- + CH_3O{-}P(=O)(OCH_3){-}OCH_3 \longrightarrow H^{18}O{-}P(=O)(OCH_3){-}OCH_3 + CH_3OH$$

As the phosphorus atom is larger than the carbon atom, steric effects in reactions at a phosphorus atom might be expected to be smaller. However, in phosphorus esters containing bulky alkyl groups marked decreases in reaction rate are observed. For example, the rate of hydrolysis of (3.97) is some five hundred times faster than that of (3.98), although these reactions are thought to follow an addition–elimination mechanism, where steric effects might be more important than in a direct substitution.

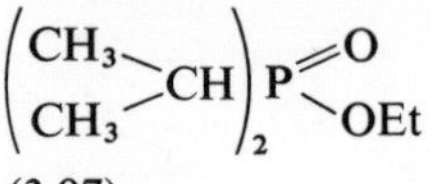

$$((CH_3)_2CH)_2P(=O)OEt \qquad t\text{-}Bu_2P(=O)OEt$$

(3.97) 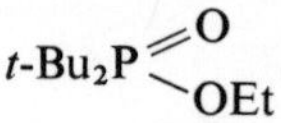 (3.98)

Electronic effects in phosphoryl compounds are not so well understood as those in carbonyl compounds (but see p. 36), partly because they have been studied less, and partly because of the involvement of d-orbitals.

Examples of reactions which follow a direct displacement mechanism are halide displacements by hydroxyl groups and many phosphorus-ester hydrolyses.

(b) *Reactions which do not involve a central phosphorus atom.* These are reactions in which the phosphorus atom, although not involved at the reaction site, modifies its reactivity. Most of the evidence for d_π–p_π overlap has been provided by reactions of this type (Chapter 1, p. 37), particularly in cases involving pentavalent tetrahedral phosphorus.

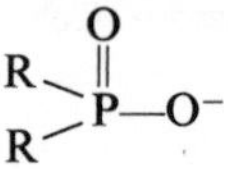

$$R_2P(=O){-}O^-$$

(3.99)

$$Me_3C{-}CH_2{-}O{-}P(=O)(Me){-}OR \longrightarrow Me_3C{-}C^+H_2 + O^-{-}P(=O)(Me){-}OR$$

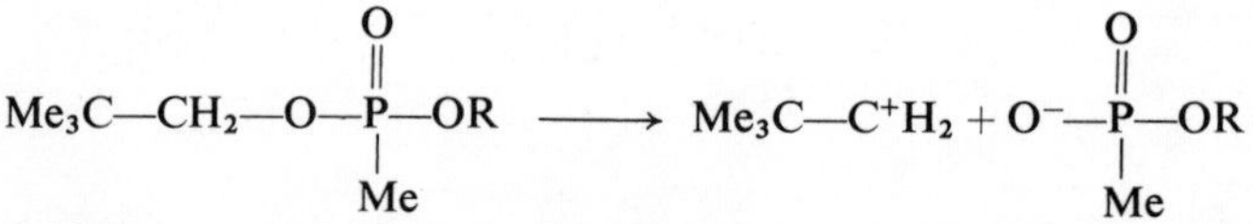

(3.100)

(i) Phosphoryl-stabilized leaving groups. Because d_π–p_π interactions disperse the negative charge, the oxy-anion (3.99) frequently occurs as a leaving group in nucleophilic substitutions (in both S_N1 and S_N2 mechanisms). For example, the phosphonate ester (3.100) appears to hydrolyse via an S_N1 mechanism, while trimethyl phosphate (3.101) reacts with aniline by a bimolecular route, yet both involve a phosphorus oxy-anion as the leaving group. These reactions provide a further pathway for the hydrolysis of phosphorus esters (see p. 107), many of

$$PhNH_2 + MeO{-}P(O)(OMe)_2 \; (3.101) \longrightarrow Ph{-}NH_2^+{-}Me + O^-{-}P(O)(OMe)_2 \longrightarrow PhNHMe + HOP(O)(OMe)_2$$

which do in fact hydrolyse by displacement at a carbon atom (cf. the hydrolysis of nitrobenzoate esters and the elimination–addition mechanism, p. 109).

Benzyl protecting groups used in nucleoside phosphorylation (see Chapter 5) are removed from the phosphorylated product (3.102) by refluxing with sodium iodide in acetone, a reaction which involves the nucleophilic displacement of a phosphorus oxy-anion. Trimethyl phosphate (3.103) acts as a mild methylating agent for phenols via a similar mechanism.

$$PhCH_2{-}O{-}P(O)(OR)(OCH_2Ph) + I^- \; (3.102) \longrightarrow O^-{-}P(O)(OR)(OCH_2Ph) + PhCH_2I$$

$$PhO^- + MeO{-}P(O)(OMe){-}OMe \; (3.103) \longrightarrow PhOMe + O^-{-}P(O)(OMe){-}OMe$$

This type of reaction is also important in terpene biosynthesis as a method of cyclization, and geraniol phosphate (3.104) has been shown to give a reasonable yield of limonene (3.105) *in vitro*.

$$\text{(3.104)} \longrightarrow \text{carbocation} + O^-{-}P(O)(OR)_2 \longrightarrow \text{(3.105)}$$

(3.104) (3.105)

The activation of multiple bonds by a phosphoryl substituent is analogous. Nucleophilic attack is facilitated by the stabilization of negative charge in the intermediate (3.106). This activation is so effective that sodium hydroxide and even water can be added to those double and triple bonds which have a phosphoryl substituent.

$$R_2P(=O)-CH{=}CH-R + {}^-OEt \longrightarrow R_2P(=O)-C^-H-CHR-OEt \;(3.106) \xrightarrow{H^+} R_2P(=O)-CH_2-CHR-OEt$$

(ii) Phosphoryl-stabilized carbanions. The potential stabilization of a negative charge on a carbon atom with a phosphoryl substituent leads to greatly increased acidity of any hydrogen atom attached to that carbon. Both phosphine oxides (3.107; R = alkyl or aryl) and phosphonates (3.107; R = *O*-alkyl or *O*-aryl) will form anions (3.108) on treatment with a variety of bases. These anions are normally stable under anhydrous oxygen-free conditions and will react with a wide range of electrophiles (see Scheme **3.1**).

$$\underset{(3.107)}{R_2P(=O)-CH_2-R'} \xrightarrow{t\text{-BuO}^-K^+} \underset{(3.108)}{R_2P(=O)-C^-H-R'\;K^+}$$

$$R_2P(=O)-C^-H-R'\;K^+ \xrightarrow{MeI} R_2P(=O)-CH(Me)-R' + KI$$

$$R_2P(=O)-C^-H-R'\;K^+ \xrightarrow{D_2O} R_2P(=O)-CHD-R' + KOD$$

$$R_2P(=O)-C^-H-R'\;K^+ \xrightarrow{PhCOOEt} R_2P(=O)-CH(CO-Ph)-R' + KOEt$$

$$R_2P(=O)-C^-H-R'\;K^+ \xrightarrow{R-CH\!-\!CHR''\ (\text{epoxide, }O)} R_2P(=O)-CH(CHR-CHR''-O^-\;K^+)-R'$$

Scheme 3.1

Probably the most important of these reactions is that between phosphonate carbanions (3.108; R = *O*-alkyl) and aldehydes and ketones; this leads to olefins via the cyclic elimination (3.109) of a phosphate anion.

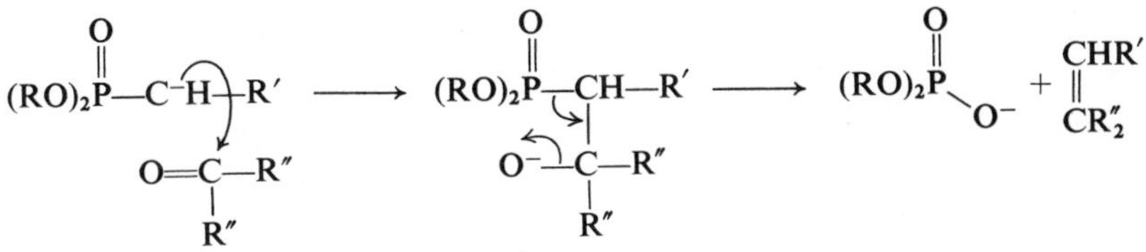

$$(RO)_2\overset{O}{\overset{\|}{P}}-C-H-R' + O{=}C(R'')-R'' \longrightarrow (RO)_2\overset{O}{\overset{\|}{P}}-CH(R')-C(R'')_2-O^- \longrightarrow (RO)_2\overset{O}{\overset{\|}{P}}-O^- + CHR'{=}CR''_2$$

(3.109)

This has been used as an alternative to the Wittig reaction (p. 142) in many cases and has the advantage both of cheaper starting materials and a more reactive carbanion; however, the stereochemistry of the olefin produced is invariably *trans*. Pommer has applied this method to the synthesis of both vitamin A (3.110) and carotenoids (3.111).

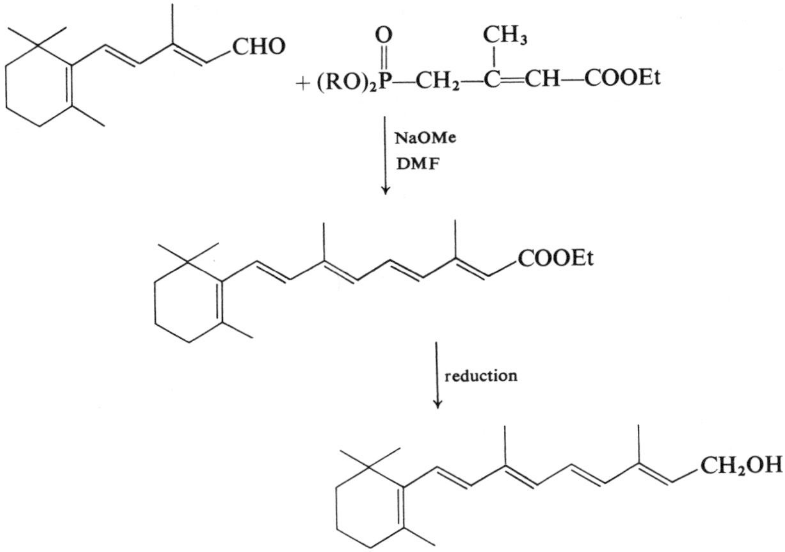

CHO + $(RO)_2\overset{O}{\overset{\|}{P}}-CH_2-\overset{CH_3}{\overset{|}{C}}{=}CH-COOEt$

NaOMe
DMF

COOEt

reduction

CH_2OH

(3.110)

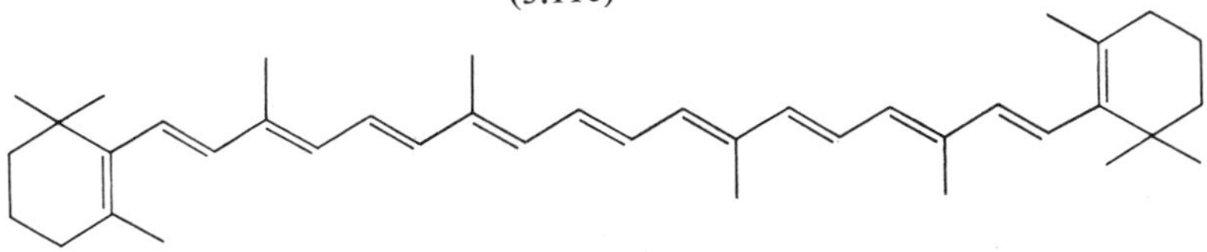

(3.111) β-carotene

The phosphoryl group itself shows some nucleophilic character at the oxygen atom, for example, the cyclization of (3.112). However this is rare, presumably because the very effective d_π–p_π overlap reduces the negative charge on the oxygen atom (see p. 132, Chapter 1).

$$(RO)_2P(=O){-}O{-}CH_2{-}CH_2{-}X \longrightarrow \left[(RO)_2P^+(OCH_2CH_2O)\right]X^- \longrightarrow RO{-}P(=O)(OCH_2CH_2O) + RX$$

(3.112)

Thio-analogues appear to be more effective in reactions of this type, and react readily with ethyl iodide. This has been cited as evidence for poorer d_π–p_π overlap in P═S as compared with P═O bonds (see p. 32).

$$(EtO)_2P(=S)Et + CH_3CH_2{-}I \longrightarrow \left[(EtO)_2P^+(SEt)Et\right]I^- \longrightarrow EtO{-}P(=O)(SEt)Et$$

3.3 Phosphazenes

The phosphazenes, or phosphonitrilics, are compounds which possess a number of repeating >P═N— units, and can be cyclic (e.g. 3.113) or linear (3.114). Compounds possessing only one such unit are variously called phosphine imides, iminophosphines, etc. (3.115), and will be discussed in Chapter 4 (p. 162).

$(R_2PN)_3$ (3.113)

$\left(R_2P{=}N\right)_n$ (3.114)

$R_3P{=}N{-}R'$ (3.115) phosphine imide

The preparation of phosphonitrilic halides (3.113 and 3.114; R = halogen) from amines and phosphorus pentahalides has already been discussed (p. 90), and most other phosphazenes are prepared from these halides by substitution reactions (p. 122).

Phosphazenes were prepared as early as 1834 by Rose and Liebig, and have emerged as a potentially useful group of compounds. Cyclic members of the series (e.g. 3.113) can be converted, by heating to around 300 °C, into a mixture of linear and cross-linked polymers of extreme thermal stability (some will resist temperatures of 600 °C). The polymers derived from the phosphonitrilic halides have excellent elastic properties, but have the disadvantage of being hydrolysed by water. The latter has been overcome to some extent by substituting other functional groups for the halogens, but usually the properties of the polymer are also affected.

Polymers with molecular weights as high as 2000000 have been prepared and used as polyelectrolytes, heat-resistant coatings, adhesives, additives for lubricating oils and flame proofing of textiles.

.3.1 *Structures*

The bonding of the cyclic phosphazenes, with their potential aromatic character, has been discussed in Chapter 1. X-ray crystallography has shown the six-membered ring (3.116) to be almost planar with approximately tetrahedral phosphorus atoms and equivalent P—N bonds; while the eight-membered ring of the tetramer can apparently exist in individually isolable boat (3.117) and chair (3.118) forms. Larger rings have not been studied in sufficient detail for their detailed structure to be known, but they are probably either puckered like carbon macrocycles, or planar reentrant (e.g. 3.119).

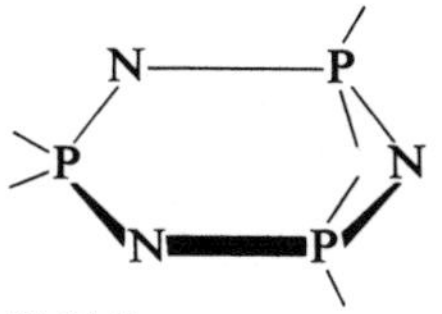

(3.116)

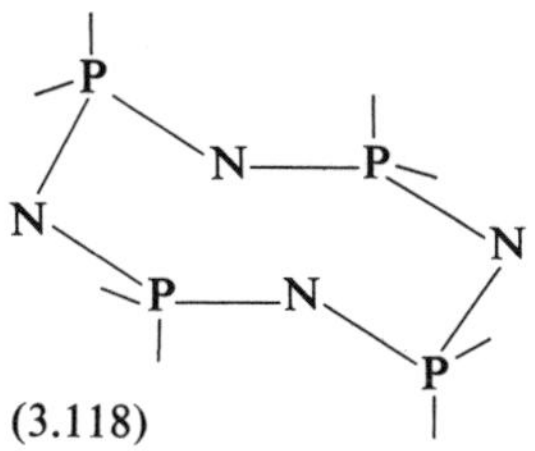

(3.118)

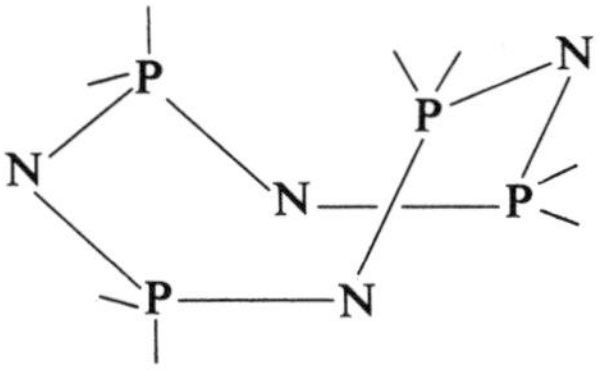

(3.117)

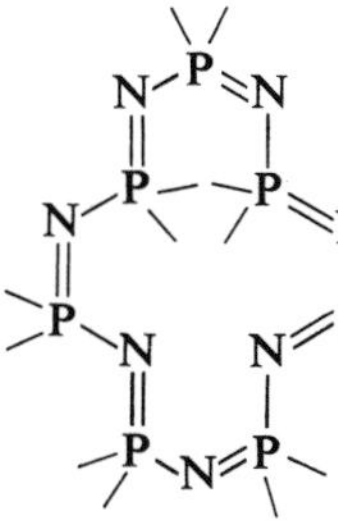

(3.119)

Cis and *trans* isomers have been isolated in a number of cases where each phosphorus atom has two different substituents, and the possible number of these isomers increases rapidly with the complexity of substituents and the ring size.

Linear molecules are also known, and exist in both covalent (3.120) and ionic (3.121) forms.

$$Cl\left[-PCl_2{=}N-\right]_n-PCl_4$$

(3.120)

$$Cl\left[-PCl_2{=}N-\right]_n-P^{+}Cl_3 \quad P^{-}Cl_6$$

(3.121; *n* is between 1 and 20)

3.3.2 *Substitution reactions*

Most phosphazene reactions fall into this category. The phosphonitrilic halides are all hydrolysed by water, although much more slowly than the simple phosphorus halides (see p. 85) (the phosphonitrilic chloride (3.113; R = Cl) can even be steam-distilled). The initial reaction involves substitution of all the halogen atoms to give a mixture of the tautomers (3.122a and b). Further hydrolysis leads to phosphoric acid and ammonia. Depending on the phosphazene halide involved, the necessary reaction conditions vary from wet ether at room temperature to hot sodium hydroxide. The equilibrium position is of interest since it involves competition between the strength of the P═O bond in (3.122b) and the strength of the P═N bond plus some aromatic stabilization in (3.122a).

$$(NPCl_2)_3 \xrightarrow{6H_2O} [NP(OH)_2]_3 \rightleftharpoons [HNPO(OH)]_3$$

(3.122a) (3.122b)

$$\downarrow 6H_2O$$

$$3H_3PO_4 + 3NH_3$$

Alcohols, both in the presence of a tertiary base and as their alcoholates, will similarly displace halogens to give alkoxyphosphazenes (3.123). Aminophosphazenes (3.124) can likewise be prepared by the reaction of primary or secondary amines with phosphonitrilic halides. The aziridyl-substituted phosphazene (3.124; $R_2N =$ aziridyl) has been used as a house-fly sterilant.

$$(NPCl_2)_3 \xrightarrow[\text{or } RO^-Na^+,\ 100\,^\circ C]{ROH/R'_3N} [NP(OR)_2]_3$$

(3.123)

$[NP(NR_2)_2]_3$

(3.124)

Grignard reagents have been used in attempts to make alkyl- and aryl-substituted phosphazenes, but the Friedel–Crafts reaction has been much more effective in the aryl case.

The mechanism of substitution appears to be mainly S_N1, although a few examples of apparently S_N2 reactions are known. The orientation of the substitutions appears to be geminal when the attacking nucleophile is electron withdrawing, but at different phosphorus atoms when it is electron donating. Some examples of substitution reactions are given in Table 22.

3.3.3 *Rearrangement reactions*

Rearrangement reactions, e.g. (3.125) to (3.126), which apparently depend on the strength of the P=O bond, are also known and suggest that any aromatic stabilization of the phosphazene ring is small, since it will be lost in the formation of (3.126).

$$[NP(OR)_2]_3 \xrightarrow{200\,^\circ C} [RN{-}P(O)(OR)]_3$$

(3.125) (3.126)

Phosphazenes will react with sulphur trioxide and aluminium trichloride, but in general are only very weakly basic unless strongly electron-donating substituents are present in the ring.

Table 22 Substitution reactions of phosphonitrilic halides

Reactant	Reagents / conditions	Product	Yield / properties
Cl, Cl, Cl, Cl, Cl, Cl (P=N ring: P, N, P, N, P, N)	6BuOH → pyridine 0 °C	BuO, BuO, OBu, OBu, BuO, OBu (P=N ring)	68 % b.p. 162 °C at 0·01 mm
Cl, Cl, Cl, Cl, Cl, Cl (P=N ring)	12 (piperidide N⁻) Na^+ → benzene reflux	six piperidino N groups (P=N ring)	80 % m.p. 231 °C
Cl, Cl, Cl, Cl, Cl, Cl (P=N ring)	6KSCN → acetone 2 h reflux	NCS, NCS, SCN, SCN, NCS, SCN (P=N ring)	85 % m.p. 41 °C

Problems

3.1 Explain why optically active phosphines are racemized by the presence of traces of bromine, but reaction with bromine in the presence of water produces optically active phosphine oxides.
[Horner and Winkler (1964).]

3.2 Most pentacoordinate phosphorus compounds are trigonal bipyramidal in structure, and the actual arrangement of groups either *apically*, or *equatorially*, appears to be controlled by factors discussed on pp. 110–13. In view of this discussion, suggest the most stable structure, or structures, for each of the following:

$(MeO)_3P$ (O, O, Me, Me) (3.127)

$(MeO)_3P$ (O, Me, COMe, Ph) (3.128)

$(MeO)_2P$ (Ph, O, Me, Me, Ph) (3.129)

F_4PMe
F_2PMe_3
(3.130)

3.3 The relative rates of alkaline hydrolysis of the esters (3.131, 3.132 and 3.133) are as follows:

(3.131) very much faster than (3.132)

about equal to (3.133)

Explain.
[Hawes and Trippet (1968).]

3.4 Dimethyl methylphosphonate (3.134; R = Me) hydrolyses at the same rate as dineopentyl methylphosphonate (3.134; R = neopentyl) in acid solution. The diisopropyl ester (3.134; R = isopropyl) hydrolyses some twenty-five times faster under the same conditions. Isopentene (3.135) is found among the products of hydrolysis of the neopentyl phosphonate (3.134; R = neopentyl). Explain.

$(RO)_2P(O)Me \longrightarrow (RO)(HO)P(O)Me + ROH$ $\qquad$ $Me_2C{=}CHMe$

(3.134) $\qquad$ (3.135)

3.5 The cyclic phosphinate (3.136) undergoes alkaline hydrolysis at approximately the same rate as its acyclic analogue ethyl diethylphosphinate (3.137), while the cyclic phosphate (3.138) hydrolyses at least 10^6 times faster than (3.139). Explain.

(3.136) $\qquad$ (3.137) $Et_2P(O)OEt$ $\qquad$ (3.138) $\qquad$ (3.139) $(EtO)_2P(O)OEt$

[See p. 110 and then read Westheimer (1968).]

3.6 Suggest reasonable mechanisms for the following reactions:

(a) $Ph_3PBr_2 + BuOBu \longrightarrow Ph_3PO + 2BuBr$

[Anderson and Freenor (1964).]

(b) $(RO)_2P(O){-}CH(OH){-}CCl_3 \xrightarrow{\text{base}} (RO)_2P(O){-}O{-}CH{=}CCl_2$

[Barthel, Alexander, Giang and Hall (1955).]

(c) $(RO)_2P(=O)-C(=O)-Me \xrightarrow{NaCN} (RO)_2P(=O)-O-CH(CN)-Me$

[Hall, Stephens and Drysdale (1957).]

(d) $(RO)(Cl)P(=O)-CH_2-CH(Me)-COOR \longrightarrow$ cyclic $ROP(=O)-CH_2-CHMe-C(=O)-O$ (ring)

[Petrov and Neĭmyesheva (1959, 1960).]

(e) $(RNH)(MeO)P(=S)O^- + CH_2-CH_2$ (epoxide, O) $\longrightarrow (RNH)(MeO)P(=O)O^- + CH_2-CH_2$ (episulfide, S)

[Hamer (1968).]

(f) cyclohexene $+ (EtO)_2P(=O)SCl \longrightarrow$ trans-2-chlorocyclohexyl $SP(O)(OEt)_2$ (Cl, $SP(OEt)_2$ with P=O) $\xrightarrow{AcOH/NaOAc}$ trans-2-acetoxycyclohexyl SAc (OAc, SAc)

[Bochwic and Frankowski (1968).]

(g) $Ph_2P(=O)-CH_2-Ph \xrightarrow[H_2O]{h\nu} Ph-P(=O)(OH)-CHPh_2 + Ph_2P(=O)-CH(OH)-Ph$

[Regitz, Anschütz, Bartz and Liedhegener (1968).]

Chapter 4
The Pentavalent State – Compounds Derived from Phosphonium Salts

4.1 Phosphonium salts

Phosphonium salts ($R_4P^+X^-$) are ionic representatives of phosphorus(V) compounds (see p. 85) and are typically high-melting-point crystalline solids. The reactions involved in the preparation of these salts have already been discussed in Chapter 2, so only the main points will be described here.

Alkyl and aryl tertiary phosphines *quaternize* with primary alkyl halides, the rate of reaction increasing from chloride to iodide (see Chapter 2, p. 53). With lower alkyl halides the reaction occurs readily, but the affinity decreases

$$R_3P: + R'X \longrightarrow R_3P^+R' \quad X^-$$

rather rapidly with increasing size of the alkyl group, and alkyl halides high in the homologous series may require reaction in sealed tubes at quite high temperatures. Polar solvents, for example nitromethane and acetonitrile, also facilitate the reaction.

Owing to their strongly nucleophilic character, tertiary phosphines generally react with alkyl halides by an S_N2 mechanism (see p. 53). Complications may arise if the halogen compound contains strongly electron-withdrawing groups which polarize the C—X bond in the sense $C^{\delta-}—X^{\delta+}$ and simultaneously stabilize the carbanion which remains after removal of the activated halogen. This then favours nucleophilic attack of the phosphine group on the halogen atom (see p. 56), but the product may not differ from the product resulting from a direct S_N2 displacement at the carbon atom (see p. 57).

$+ \; Ph_3P^+H \quad Cl^-$

DMF

P^+Ph_3

$CH_2 \quad Cl^-$

(4.1)

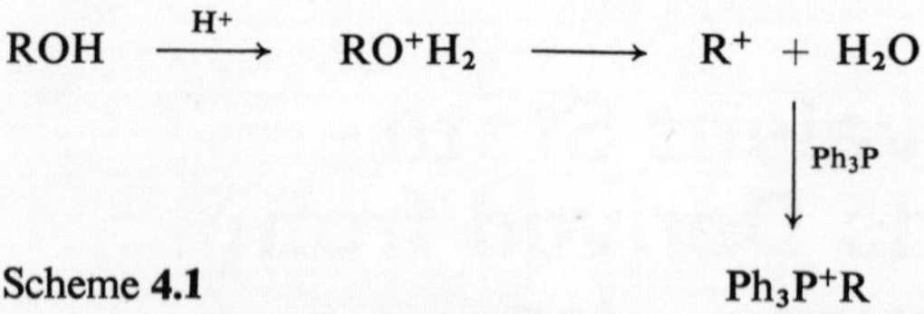

$$ROH \xrightarrow{H^+} RO^+H_2 \longrightarrow R^+ + H_2O$$

$$R^+ \xrightarrow{Ph_3P} Ph_3P^+R$$

Scheme **4.1**

Pommer and his co-workers have prepared phosphonium salts (4.1), which are useful in the carotenoid field, by the novel method of adding triphenylphosphonium halides to polyenes. High yields of phosphonium salts in the vitamin A series have also been obtained from the reaction of alcohols with tertiary phosphines and acid, probably via the ionic mechanism (Scheme **4.1**) (see p. 54). Support for this mechanism comes from the reaction of triphenylphosphine with β-ionylidene ethanol (4.2) and its isomer, 9-vinyl-β-ionol (4.3), which give the same β-ionylideneethyltriphenylphosphonium salt (4.4).

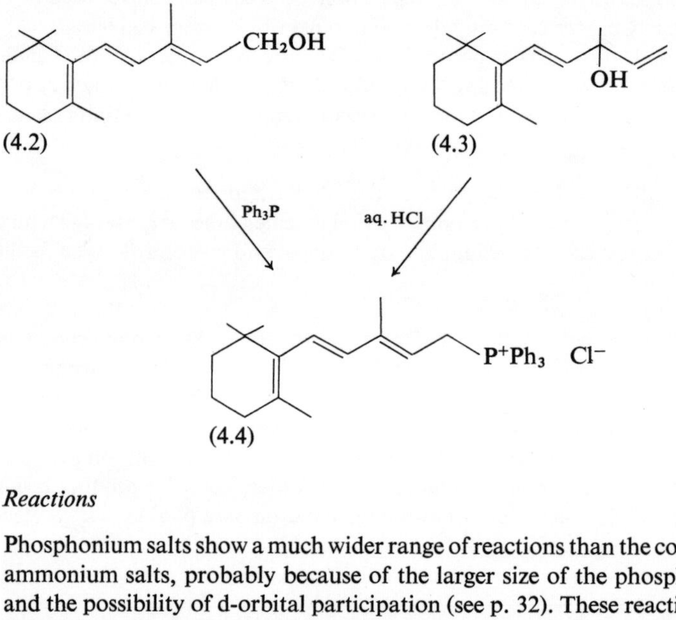

CH₂OH

(4.2)

OH

(4.3)

Ph_3P aq. HCl

P^+Ph_3 Cl^-

(4.4)

4.1.1 *Reactions*

Phosphonium salts show a much wider range of reactions than the corresponding ammonium salts, probably because of the larger size of the phosphorus atom, and the possibility of d-orbital participation (see p. 32). These reactions are best divided into four types on the basis of the site of initial attack on the phosphonium salt (4.5) by the reagent.

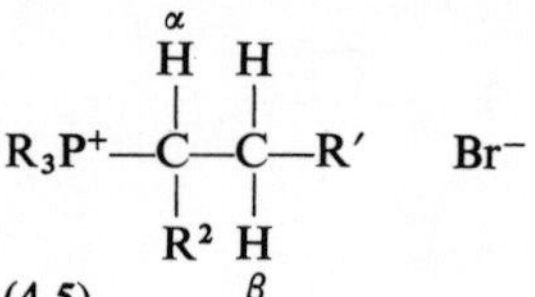

$$R_3P^+—\overset{\overset{\alpha}{H}}{\underset{R^2}{C}}—\overset{H}{\underset{\underset{\beta}{H}}{C}}—R' \quad Br^-$$

(4.5)

(a) *Attack at the α-carbon atom.* Phosphonium salts undergo decomposition on heating to give tertiary phosphines (4.7) and organic halides (4.6) (this decomposition is the reverse of quaternization).

$$X^- + R'CH_2{-}P^+R_3 \longrightarrow R'CH_2X + R_3P$$

(4.6) (4.7)

It was originally reported that the pyrolysis of tetraalkylphosphonium salts gave olefins (as happens with ammonium salts), but Ingold later disproved this. However, decomposition of long-chain phosphonium salts can lead to the simultaneous formation of varying amounts of olefins by attack on the β-hydrogen atom. Quaternary salts with two or more different substituents on the phosphorus atom can cleave in several ways. The general order of cleavage is

ethyl < benzyl < methyl < propyl < isoamyl < phenyl

but the differences are small, and mixtures are usually obtained.

A similar type of reaction has been observed as a side reaction occurring concurrently with ylid formation (see p. 136). For example, when diphenoxymethyltriphenylphosphonium chloride (4.8) is treated with phenyllithium, some 7 per cent of the product is derived from nucleophilic attack at the α-carbon atom by the base.

$$Ph_3P^+{-}CH(OPh){-}OPh\ Cl^- + Ph^-\ Li^+ \longrightarrow Ph_3P^+{-}CH(Ph){-}OPh\ Cl^-$$

(4.8)

(b) *Attack at the phosphorus atom.* As might be expected, nucleophilic reagents commonly attack a phosphonium salt at the positively charged phosphorus atom.

Aqueous alkali reacts with phosphonium salts in two different ways. Hydroxide can act as a base and remove an α-hydrogen atom, if these are sufficiently acidic, to give a phosphonium ylid (see p. 136). Alternatively, hydroxide can carry out a nucleophilic displacement at the phosphorus atom to give a phosphine oxide (4.9) and a hydrocarbon (4.10).

$$R_4P^+\ OH^- \longrightarrow R_3P{=}O + RH$$

(4.9) (4.10)

This hydrolysis reaction shows third-order kinetics with rate proportional to [phosphonium salt] × $[OH^-]^2$, and the generally accepted pathway is shown in Scheme **4.2**.

$$R_3P^+\text{—}R' + OH^- \underset{}{\overset{\text{fast}}{\rightleftharpoons}} R_3P(OH)\text{—}R' \overset{\text{fast}}{\rightleftharpoons} R_3P(O^-)\text{—}R' \overset{\text{slow}}{\rightleftharpoons} R_3P{=}O + R'^- \xrightarrow{H_2O} R'H$$

Scheme **4.2**

The nature of the group R′ lost from phosphonium salts containing more than one type of substituent depends on the stability of the group as an anion, that is, groups are increasingly readily displaced in the order

$$CH_3\text{—}CH_2 < PhCH_2CH_2 < CH_3 < Ph < PhCH_2 < p\text{-}NO_2C_6H_4CH_2$$

This means that tetraalkylphosphonium salts are extremely difficult to hydrolyse; for example, tetraethylphosphonium iodide (4.11) refluxed with 5N aqueous sodium hydroxide for a week shows only 10 per cent hydrolysis, while the phosphonium salt (4.12) is 100 per cent hydrolysed in an hour under the same conditions.

$$\underset{(4.11)}{Et_4P^+I^-} \xrightarrow{OH^-} Et_3P{=}O + EtH$$

$$\underset{(4.12)}{Ph_3P^+\text{—}CH_2PhI^-} \xrightarrow{OH^-} Ph_3P{=}O + PhCH_3$$

Simple phosphonium salts are hydrolysed with inversion of the configuration at the phosphorus atom (but see p. 133). Horner prepared optically active phosphonium salts (e.g. 4.13) and determined the stereochemistry of their various modes of decomposition (Scheme **4.3**).

$$PhCH_3 + O{=}P(Ph)(Et)(Me) \; (4.16)$$

$$(Ph)(Et)(Me)P{=}O \; (4.15)$$

$$(4.13) \xrightarrow[\text{inversion}]{OH^-} (4.16)$$

$$(4.14) \xrightarrow[\text{retention}]{H_2O_2} (4.15)$$

$$\underset{(4.13)}{(Ph)(Et)(Me)P^+\text{—}CH_2\text{—}Ph} \underset{\text{retention}}{\overset{\text{electrolysis}}{\rightleftharpoons}} \underset{(4.14)}{(Ph)(Et)(Me)P:}$$

Scheme **4.3**

He showed that electrolysis to give the phosphine (4.14), followed by oxidation, gave the phosphine oxide (4.15); this is of opposite sign of rotation to the oxide (4.16) obtained by alkaline hydrolysis of the phosphonium salt (4.13). These results were interpreted as involving inversion of the configuration at the phosphorus atom during hydrolysis, as shown in Scheme **4.3**.

The detailed mechanism of hydrolysis is thought to involve a phosphorus(V) intermediate (4.17) with an sp^3d-hybridized central atom (i.e. trigonal-bipyramidal structure). Because of the geometry of a trigonal bipyramid, an inversion of configuration must involve either apical–apical elimination (4.18), or equatorial–equatorial elimination (4.19). In fact the elimination is

(4.18) (4.19) (4.17)

thought to be collinear (i.e. apical–apical). Of course, if this trigonal-bipyramidal intermediate has a sufficient lifespan, it may racemize anyway by a process of pseudo-rotation (see p. 31); this obviously does not happen when inversion occurs.

Recent work has concentrated on various means of restricting the geometry of the trigonal bipyramid. Trippett has used a four-membered ring to accomplish this; he reasons that the differences in angle between equatorial–apical (90°) and equatorial–equatorial (120°) bonds in the trigonal-bipyramidal intermediate would ensure that the four-membered ring bridged apical and equatorial positions (4.20), rather than two equatorial positions (4.21).

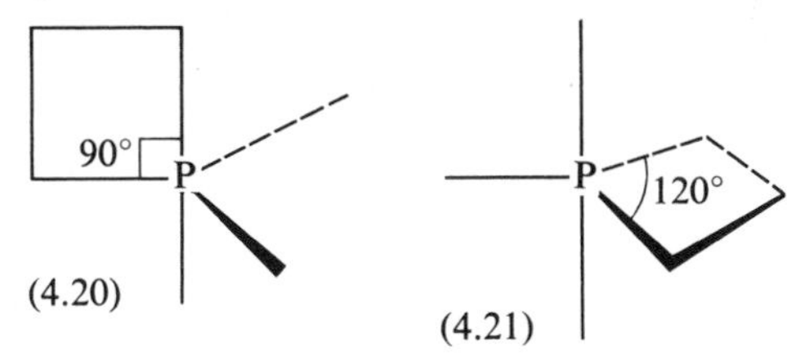

(4.20) (4.21)

Alkaline hydrolysis of the phosphetanium salt (4.22) would be expected, since the phenyl group is the most stable as an anion, to lose this and form the phos-

(4.22) (4.23)

phine oxide (4.23) containing an intact four-membered ring. However, the product obtained from alkaline hydrolysis is the five-membered cyclic phosphine oxide (4.24).

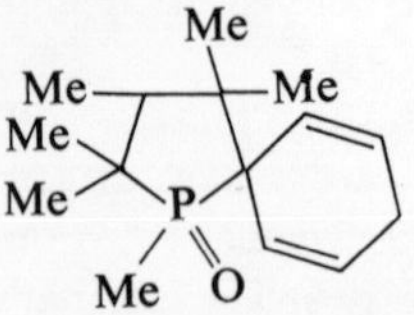

(4.24) 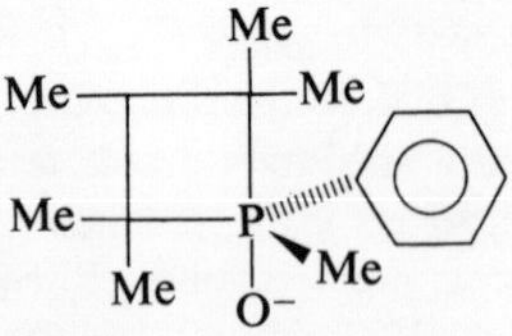 (4.25)

Because of the constraints imposed by the four-membered ring, the trigonal-bipyramidal intermediate in the hydrolysis of (4.22) can only have the geometry (4.25), with the ring in an equatorial–apical position; therefore the phenyl group and the oxy-anion cannot both become apical and collinear. The intermediate eliminates the group now in the apical position (4.26), even though this is not the group most stable as the anion, because this process is the most favourable energetically. The eliminated anion then migrates to the phenyl ring, which can delocalize its charge to some extent, and protonation finally produces (4.27).

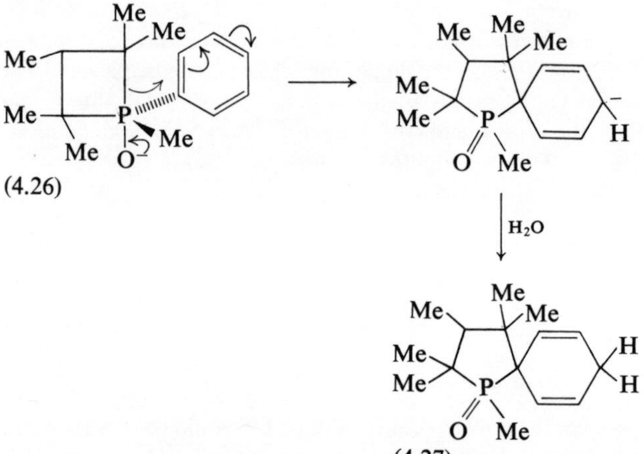

(4.26)

(4.27)

Other salts containing four- and five-membered rings (e.g. 4.28 and 4.29) give apparently abnormal products for the same reasons (see Problems, p. 170).

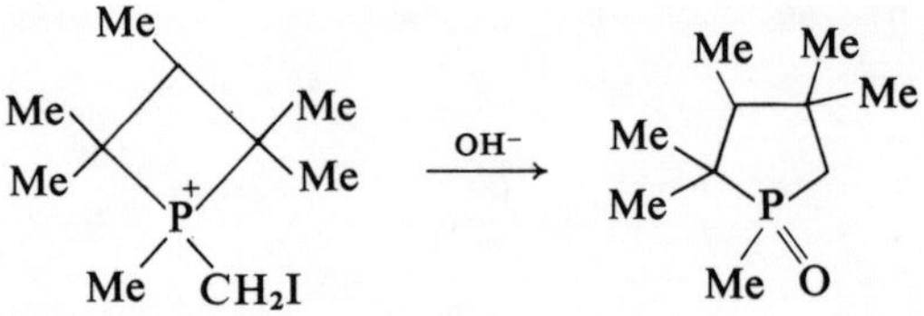

(4.28)

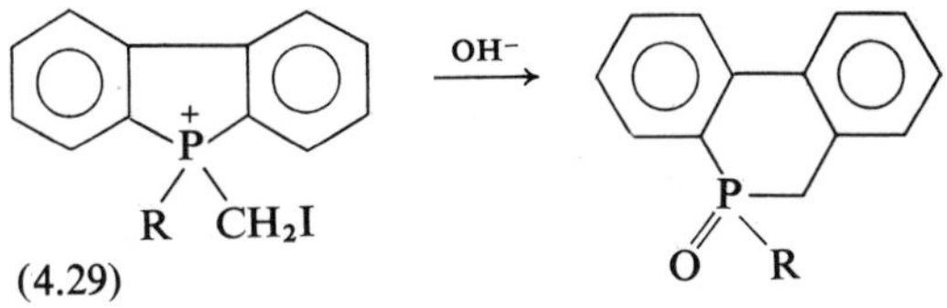

(4.29)

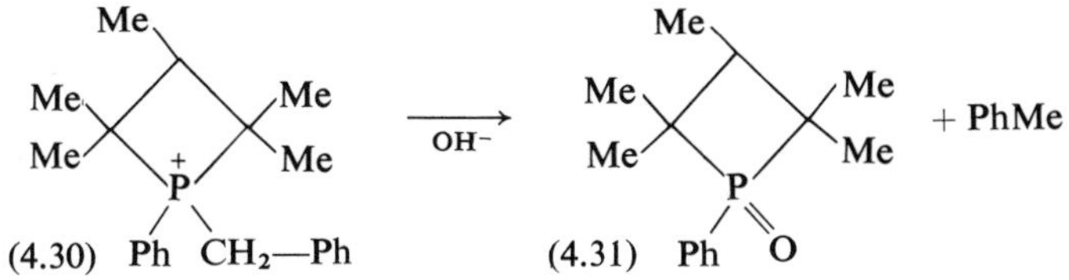

(4.30) (4.31)

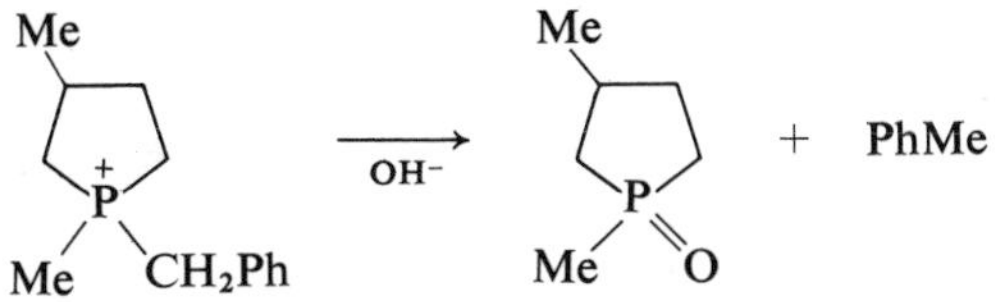

$+ PhMe$

(4.32) (4.33)

$+ PhMe$

Hydrolysis of the salts (4.30 and 4.32) gives the phosphine oxides (4.31 and 4.33) by loss of a benzyl group, presumably because the energy of the abnormal route described above is too high. However, unlike normal phosphonium salt hydrolyses, these were found to take place with retention of configuration. The same restrictions (i.e. apical–equatorial ring bridging) as in the case of (4.25) apply to the trigonal-bipyramidal intermediates derived from (4.30) and (4.32). However, in (4.34) and (4.35) there is no alternative pathway available, and so a benzyl group is lost from an equatorial position to give the expected phosphine oxide; but, since there has been apical–equatorial elimination, the configuration is retained.

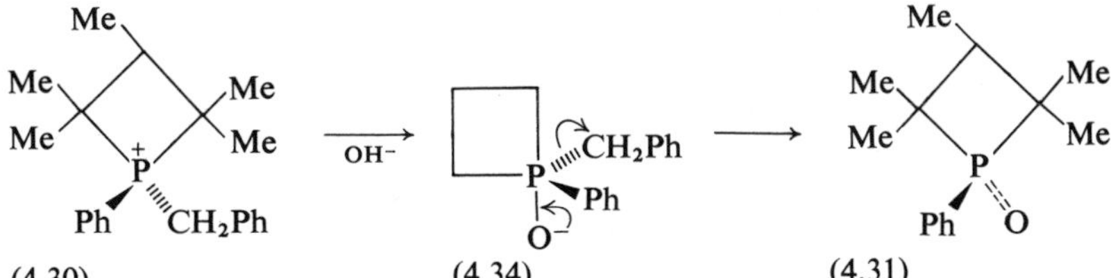

(4.30) (4.34) (4.31)

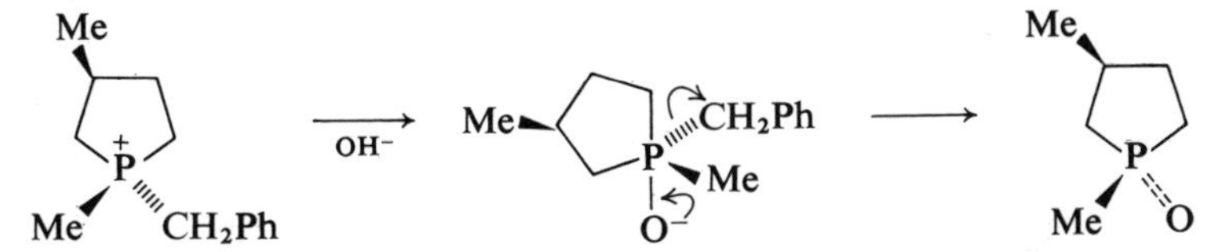

(4.32) (4.35) (4.33)

$$R_4P^+ \; O^-CH_2CH_3 \xrightarrow{>120\,°C} R_3P{=}O + RH$$
(4.36)

Alkoxide ions (RO^-) can attack the phosphorus atom of phosphonium salts in a similar way to hydroxide ions. However, this reaction, known as alcoholysis, is much slower than hydrolysis. Phosphonium alkoxides (4.36) generally decompose above 120 °C to give phosphine oxides and hydrocarbons; but, in solution, the decomposition only takes place when one of the R groups is a good leaving group, such as benzyl or $—CH_2—CO—CH_3$. The products in solution are phosphine oxide (4.40), hydrocarbon and an ether (4.39) derived from the alkoxide anion. The suggested mechanism involves loss of a benzyl group from the phosphorus(V) intermediate (4.37), followed by nucleophilic displacement at a carbon atom (4.38) by the alkoxide anion. In contrast to hydrolysis, alcoholysis almost invariably leads to racemization, possibly because the intermediate trigonal bipyramid (4.37) has a longer lifespan than the intermediate derived from hydrolysis, and so can racemize by a number of pseudo-rotations. However, it has been suggested that this racemization is due to reversible alkoxide displacement on the intermediate (4.41).

$R_3P^+—CH_2—Ph \;\; O^-CH_2CH_3$

↓

(4.37): CH_2Ph, R, R, R on P, OCH_2CH_3 → (4.38): $R_3P^+—O—CH_2CH_3$ + $O^-CH_2CH_3$, + $PhCH_2^-$ $\xrightarrow{CH_3CH_2OH}$ $PhCH_3 + CH_3CH_2O^-$

(4.37) (4.38)

↓

$(CH_3CH_2)_2O + R_3P{=}O$
(4.39) (4.40)

$$CH_3CH_2O^- \;\; R_3P^+—OCH_2CH_3 \rightleftharpoons CH_3CH_2O—P^+R_3 + O^-CH_2CH_3$$
(4.41)

Metal alkyls normally react with phosphonium salts by attack at an α-hydrogen atom (see p. 136), but instances of attack at the phosphorus atom are known. For example, when methyltriphenylphosphonium bromide (4.42) is treated with methyllithium in ether, although the main product is the expected phosphonium ylid (4.43), a 26 per cent yield of benzene is also produced, presumably by the mechanism shown.

$$Ph_3P^+{-}CH_3 \xrightarrow{CH_3Li} Ph_3P^+{-}C^-H_2 + CH_4$$

(4.42) → (4.43)

$$Ph_3P^+{-}CH_3 \xrightarrow{CH_3Li} CH_3{-}P(Ph)(Ph)(Ph){-}CH_2{-}H \longrightarrow CH_3P^+(Ph_2){-}C^-H_2 + PhH \quad 26\%$$

(c) *Attack at a β-hydrogen atom.* The usual reaction of quaternary ammonium salts (4.44) with basic reagents is removal of a β-hydrogen atom and elimination to give an olefin (the Hofmann elimination). This reaction does not usually occur with phosphonium salts, because a number of more favourable alternatives are normally available. These differences have been quoted as evidence for d-orbital participation in phosphonium salts (see p. 38). However, in particularly favourable cases Hofmann elimination does take place. On heating, β-phenylethyltriphenylphosphonium hydroxide (4.45) gives mainly benzene and β-phenylethyldiphenylphosphine oxide (4.46) but some styrene and triphenylphosphine are also obtained. The introduction of a second β-phenyl substituent (4.47) makes Hofmann elimination the main pathway.

$$R_3N^+{-}CH_2{-}CH(H){-}R' \xrightarrow{:base} R_3N + CH_2{=}CH{-}R'$$

(4.44)

$$Ph_3P^+{-}CH_2{-}CH_2{-}Ph\ OH^- \xrightarrow{\text{attack at phosphorus}} Ph_2P(=O)CH_2CH_2Ph + PhH$$

(4.45) → (4.46)

$$Ph_3P^+{-}CH_2{-}CH_2{-}Ph\ OH^- \xrightarrow{\text{Hofmann elimination}} Ph_3P + CH_2{=}CH{-}Ph$$

$$Ph_3P^+{-}CH_2{-}CH(Ph)_2\ OH^- \longrightarrow Ph_3P + CH_2{=}C(Ph)_2$$

(4.47)

$$Ph_3P^+{-}CH_2CH_2OH\ OH^- \longrightarrow Ph_2P(=O)CH_2CH_2OH + PhH$$

(4.48)

$$Ph_3P^+{-}CH_2{-}CH_2{-}O{-}COCH_3\ OH^- \longrightarrow Ph_3P + CH_2{=}CH.O.COCH_3$$

(4.49)

$$Ph_3P^+—CH_2OH \quad OH^- \longrightarrow Ph_3P + CH_2{=}O$$
(4.50)

Surprisingly, β-hydroxyphosphonium hydroxides (4.48) and β-acetoxyphosphonium hydroxides (4.49) differ in that the former give phosphine oxide while the latter undergo Hofmann elimination. However, α-hydroxy salts (4.50) readily eliminate phosphine in aqueous base to give a high yield of formaldehyde.

(d) *Attack at an α-hydrogen atom.* Bases may react with phosphonium salts in a number of ways depending on the salt, the base and the conditions used, but by far the commonest reaction is the removal of a proton from a carbon atom adjacent to the phosphorus atom to give a phosphonium ylid (4.52). The formation of ylids in this way, and their stability, has been used as evidence for d_π–p_π bonding between the phosphorus atom and the adjacent carbanion. Certainly phosphorus ylids are very much more stable than their nitrogen equivalents, whereas on a purely electrostatic basis the opposite should be the case (see p. 36). If d_π–p_π bonding does take place, the ylids are best represented as a resonance hybrid of the canonical forms (4.53a and b) and both representations are widely used (see pp. 138–41, 162).

$$R_3P^+—CH_2—R' \xrightarrow{\text{base}} R_3P^+—CH^-—R'$$
(4.51) (4.52)

$$R_3P^+—C^-H—R' \longleftrightarrow R_3P{=}CH—R'$$
(4.53a) (4.53b)

The strength of base required to form an ylid from a phosphonium salt depends on the acidity of the α-hydrogen atoms and this is controlled by the substituents on the α-carbon atom. The more the substituent R′ in the salt (4.51) is able to stabilize an adjacent negative charge, the more easily will protons on the carbon atoms adjacent to it be removed, and the more stable will be the phosphonium ylid formed. For example, while methyltriphenylphosphonium bromide (4.54) will only form an ylid on treatment with metal alkyls in inert solvents under nitrogen, di(ethoxycarbonyl)methyltriphenylphosphonium bromide (4.55) is converted to a perfectly stable ylid on treatment with triethylamine. A variety of bases and the type of phosphonium salt with which they can be used is given in Table 23.

$$\overset{Br^-}{Ph_3P^+—CH_3} \xrightarrow[\text{ether},\ N_2]{\text{RLi}} Ph_3P^+—C^-H_2$$
(4.54)

$$\overset{Br^-}{Ph_3P^+—CH(COOEt)_2} \xrightarrow{Et_3N} Ph_3P^+—C^-(COOEt)_2$$
(4.55)

Table 23 Bases used in the Wittig reaction

Base	*Solvent*	*Example of phosphonium salt*
R_3N	neat	$Ph_3P^+—CH(COOR)_2$ X^-
Na_2CO_3	10% aqueous	$Ph_3P^+—CH_2—CO—Me$ X^-
Na^+OH^-	10% aqueous	$Ph_3P^+—CH_2—COOEt$ X^-
NH_3	neat	$Ph_3P^+—CHPh_2$ X^-
RO^-Na^+ e.g. MeO^-Na^+	ROH	$Ph_3P^+—CH_2Ph$ X^-
	DMF	$Ph_3P^+—CH_2.CH{=}CH.COOR$ X^-
t-BuO^-K^+	benzene THF di-n-butyl ether	$Ph_3P^+—CH_2Ph$ X^-
$Ph_3C^-Na^+$	benzene	$Ph_3P^+—alkyl$ X^-
$NaNH_2$	NH_3 liquid	$Ph_3P^+(X^-)—CH_2—CH_2—CH_2—P^+(X^-)Ph_3$
$LiNR_2$	R_2NH NH_3 liquid benzene	$Ph_3P^+—alkyl$ X^-
NaH	DMSO (CH_3SOCH_3)	$Ph_3P^+—alkyl$ X^-
	NaH in this exists as $CH_3SO—CH_2^-Na^+$	$Ph_3P^+—CH_2—Ar$ X^- Must not contain a $>C{=}O$ group.
LiPh	benzene hexane (occasionally ether, but slowly reacts with LiPh)	$Ph_3P^+—CH_2Ar$ X^- $Ph_3P^+—alkyl$ X^- Must not contain a $>C{=}O$ group.
Li alkyls e.g. LiMe LiBu	hexane pentane	$Ph_3P^+—alkyl$ X^- $Ph_3P^+—CH_2—Ar$ X^-
potassium	benzene	$Ph_3P^+—alkyl$ X^- H Ph_3P^+

The table contains some of the bases which have been used in the Wittig reaction, listed in approximate order of increasing base strength. The choice of base depends mainly on the acidity of the phosphonium salt, although other factors, such as reactivity with other functional groups, availability and precautions required, are often important.

A widely used method of ylid preparation involves the use of a solution of sodium alkoxide in the corresponding alcohol. This may be used for a wide range of ylids, although for the least stable ones the reaction involves an equilibrium between ylid (4.57) and phosphonium salt (4.56), with a very low concentration of ylid. However, for the Wittig reaction (see p. 142) this does not usually matter, since the aldehyde (or ketone) can be added to the reaction mixture and will react with the ylid (4.57) as it is formed.

$$R_3P^+{-}CH_2{-}R' + O^-R'' \rightleftharpoons R_3P^+{-}C^-H{-}R' + R''OH$$
(4.56) (4.57)

$$Ph_3P^+{-}\underset{H}{\overset{Br^-}{CH}}{-}X$$
(4.58)

LiR ↙ ↘ LiR′

$$Ph_3P^+{-}C^-H{-}X + RH + LiBr \qquad Ph_3P^+{-}C^-H_2 + R'X + LiBr$$

Phosphonium salts with α-heteroatom substituents will sometimes show abnormal reactions with bases. For example, α-bromophosphonium and α-iodophosphonium salts (4.58; X = Br or I) will react with strong bases with removal of either halogen or hydrogen; the route followed can be controlled by the choice of base.

4.2 Phosphonium ylids

The term ylid was first introduced by the father of modern organophosphorus chemistry, Georg Wittig, and refers to compounds which in their ground state have charges of opposite sign on adjacent atoms (4.59). Phosphonium ylids (4.60) are but one example of this type of compound and later in this chapter (p. 162) we will meet another, the imidophosphoranes (4.61). In both examples, the positive end of the dipole is a phosphorus atom, though ylids need not contain a phosphorus atom. However, because of their stability, phosphorus-containing ylids have tended to be studied most extensively.

$$X^+{-}Y^- \qquad >P^+{-}C^-< \qquad >P^+{-}N^-{-}$$
(4.59) (4.60) (4.61)

$$R_3P^+{-}C^-H{-}R' \longleftrightarrow R_3P{=}CH{-}R'$$
(4.53a) (4.53b)

Phosphonium ylids are really carbanions, but the carbanionic character is modified by the adjacent positive charge, and their reactivity is controlled by two factors. We have already mentioned (p. 37, Chapter 1) the possibility of d_π–p_π bonding in these compounds and, in view of this, phosphonium ylids are probably best represented by a resonance hybrid with the canonical forms (4.53a and b). The extent of d_π–p_π bonding will control the contribution from (4.53b) (since the major characteristic of ylids is their carbanion character, they are represented in this chapter by (4.53a) rather than (4.53b), but this is not intended to suggest a predominant contribution from either canonical form in any particular case).

The nature of the substituent R′ further modifies the ylid carbanion, resulting in a pattern of reactivity ranging widely from a violent reaction with air and water (e.g. when R′ = alkyl) to stability in hot alkali (e.g. 4.62).

Ph_3P^+—$C_5H_4^-$

(4.62)

Because of this large variation in chemical reactivity, ylids are normally divided into two groups: *reactive* ylids and *stable*, or *unreactive*, ylids. When the substituent R′ is strongly electron withdrawing (e.g. ester), the carbanion will be delocalized, and the ylid will be stable; however, when it is electron donating (e.g. alkyl), the carbanion will have its charge concentrated on the α-carbon atom, and the ylid will be reactive. Obviously there will be a whole spectrum of reactivity in between, and some ylids have an intermediate reactivity. Examples of stable and reactive ylids are given in Table 24.

Probably the first phosphonium ylid to be studied was that obtained by Michaelis in 1899 on treatment of phenacyltriphenylphosphonium bromide (4.63) with sodium hydroxide solution. A number of structures were suggested for this stable ylid until, in 1957, Ramirez and Dershowitz proposed the now-accepted resonance hybrid of forms (4.46a, b and c). This explained the very low frequency infrared carbonyl absorption (1529 cm^{-1}) by the contribution from (4.64c). Later, compounds of this type were shown to give acetylenes on pyrolysis, presumably again assisted by contributions from (4.64c) (cf. Wittig reaction, p. 142). It is now generally accepted that stable ylids have considerable resonance interaction with the electron-withdrawing substituents on the carbanionic carbon atom.

$$Ph—CO—CH_2—Br + Ph_3P: \longrightarrow Ph—CO—CH_2—P^+Ph_3\ \ Br^-\ \ (4.63)$$

$$\xrightarrow{NaOH\ (aq.)} \text{phosphonium ylid}$$

Table 24 Stability of ylids

most stable	$Ph_3P^+—C^-(COOR)_2$
	$Ph_3P^+—$ (cyclopentadienide)
	$Ph_3P^+—C^-H—COR$
	$Ph_3P^+—C^-H—COOR$
	$Ph_3P^+—C^-H—CN$
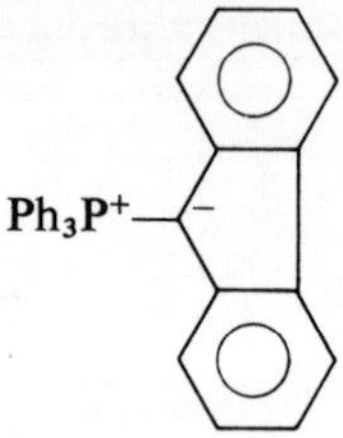	$Ph_3P^+—$ (fluorenide)
	$Ph_3P^+—C^-H—CH{=}CH—COOR$
	$Ph_3P^+—C^-(Ph)_2$
	$Ph_3P^+—C^-H—C_6H_4—NO_2$
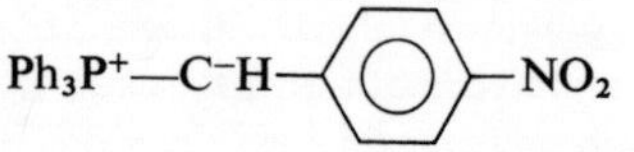	$Ph_3P^+—C^-H—Ph$
	$Ph_3P^+—C^-H—CH{=}CH_2$
	$Ph_3P^+—C^-H—C_6H_4—OMe$
	$Ph_3P^+—C^-{=}CH_2$
	$Ph_3P^+—C^-H_2$
	$Ph_3P^+—C^-H—(CH_2)_2—CH_3$
	$Ph_3P^+—C^-(CH_3)_2$
least stable	$Ph_3P^+—$ (cyclohexylidene, $^-$)

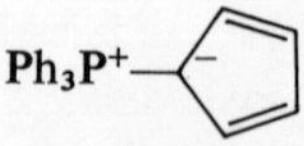

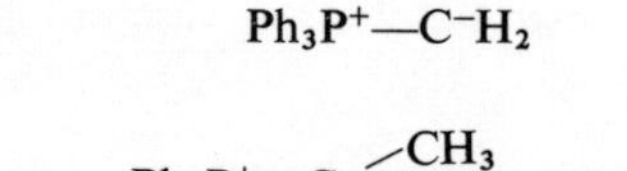

$$Ph_3P{=}CH{-}CO{-}Ph \longleftrightarrow Ph_3P^+{-}C^-H{-}CO{-}Ph \longleftrightarrow Ph_3P^+{-}CH{=}C(O^-){-}Ph$$

(4.64a) (4.64b) (4.64c)

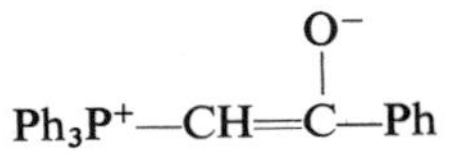

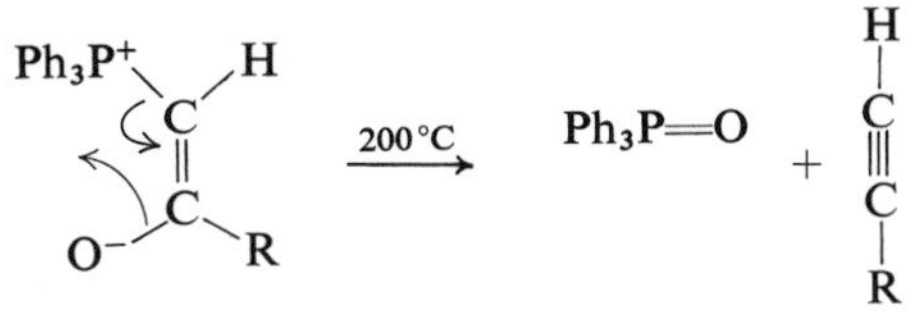

$$Ph_3P^+{-}CH{=}C(O^-){-}R \xrightarrow{200\,°C} Ph_3P{=}O + H{-}C{\equiv}C{-}R$$

Although reactive ylids like (4.53; R′ = alkyl) are not complicated by resonance with R′, their study is made difficult by their chemical reactivity. They are usually prepared, and reacted, in solution (which is normally orange-red in colour) under a nitrogen atmosphere, although they can be isolated as crystalline solids in an inert atmosphere. Because of practical difficulties, it was not until 1969 that an X-ray structure determination was carried out on a reactive ylid (methylenetriphenylphosphorane 4.65).

$$Ph_3P^+{-}C^-H_2 \longleftrightarrow Ph_3P{=}CH_2$$

(4.65a) (4.65b)

The length of the P—C ylid bond was determined as $1{\cdot}66 \times 10^{-10}$ m (1·66 Å), which is considerably less than the, presumably single, P—C bond length of $1{\cdot}85 \times 10^{-10}$ m (1·85 Å) in tertiary phosphines (R_3P). This suggests that the P—C ylid bond has a lot of double-bond character, that is, there is a large contribution from (4.65b). This double-bond character is probably due to p_π–d_π bonding (see p. 32, Chapter 1). It is interesting that the sum of Pauling's double-bond radii of phosphorus and carbon atoms modified by the Schomaker–Stevenson rule (see p. 21, Chapter 1) is $1{\cdot}665 \times 10^{-10}$ m (1·665 Å), very close to that observed for the ylid P—C bond length.

The values of bond angles at the carbanion carbon atom would give very useful information about the hybridized state of this atom. Unfortunately, X-ray crystallographic studies cannot pinpoint hydrogen atoms with any accuracy (because of their low electron density), and so this information is not available. The most reasonable state would appear to be sp^2 hybridization (4.66), rather than sp^3 (4.67), because this would lead to a maximum π-overlap with the 3d orbitals on the phosphorus atom. This suggestion is supported by ^{13}C–^{1}H coupling constants, which often give an indication of the hybridized state of the interacting carbon atom.

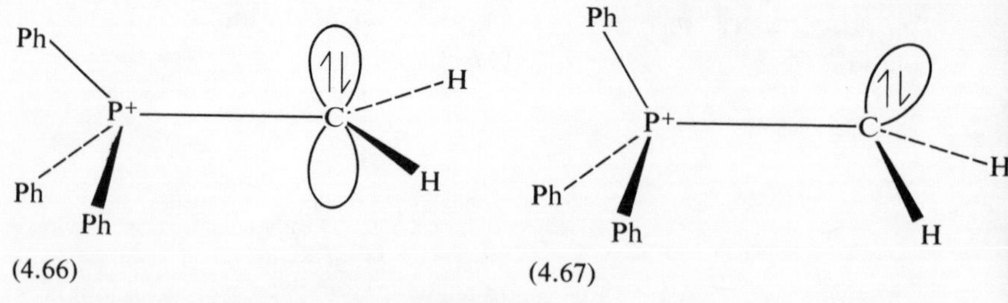

(4.66) (4.67)

So far we have only discussed the substituents at the carbanion and their effect. Substituents on the phosphorus atom also appear to be important; ylids with aryl substituents are more stable chemically than those with alkyl substituents, presumably because, in the former, d_π–p_π bonding with the phosphorus atom is in some way enhanced. However, work of this type is rather fragmentary, and a thorough, systematic study is only just beginning to appear.

Because of their carbanion character, phosphonium ylids will react with a large range of electrophiles, yet they also contain an electrophilic phosphorus atom. This centre of positive charge enables them also to react with nucleophilic reagents. In such reactions they usually behave like the corresponding phosphonium salts; in fact, the first step in many reactions of this type is protonation to give the phosphonium salt. This is exemplified by the reaction of reactive phosphonium ylids (4.68) (see p. 151) with water.

$$R_3P^+\text{—}CH^-\text{—}R' \xrightarrow{H_2O} R_3P^+\text{—}CH_2\text{—}R'$$

(4.68)

$$\downarrow OH^-$$

$$R_3P{=}O + R'CH_3$$

4.2.1 *The Wittig reaction*

Undoubtedly the most important reaction of phosphonium ylids so far discovered is that with aldehydes (or ketones) to give olefins and phosphine oxides (4.69).

$$R_3P^+\text{—}C^-R^1R^2 + R^3R^4C{=}O \longrightarrow R^1R^2C{=}CR^3R^4 + R_3P{=}O \quad (4.69)$$

This reaction, which is now universally known as the *Wittig reaction*, was originally discovered by Hans Staudinger in the 1920s and developed into a useful synthetic method by Georg Wittig in the 1950s. The reaction has several advantages; it usually proceeds under very mild conditions, the carbonyl compound may contain a wide range of functional groups (e.g. halogen, ester, acetal, ether, hydroxyl) and, unlike the Grignard method (Scheme **4.4**), there is no ambiguity in the position of the double bond being formed (Scheme **4.5**).

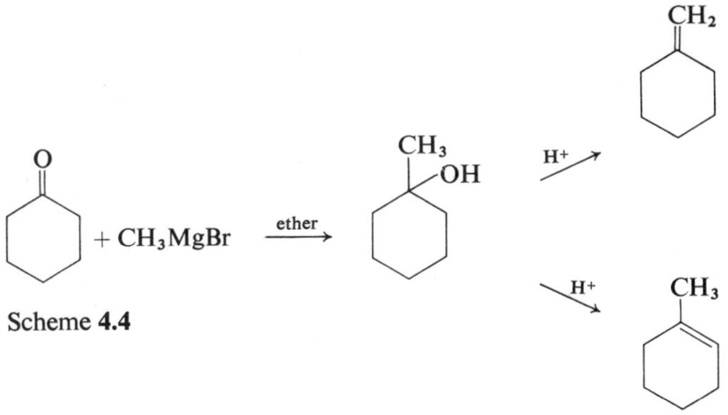

$$\text{cyclohexanone} + CH_3MgBr \xrightarrow{\text{ether}} \text{1-methylcyclohexanol} \xrightarrow{H^+} \text{methylenecyclohexane} \text{ or } \text{1-methylcyclohexene}$$

Scheme **4.4**

$$\text{cyclohexanone} + Ph_3P^+\text{—}C^-H_2 \xrightarrow[\text{reflux, 24 h}]{\text{ether}} \text{methylenecyclohexane } (70\%) + Ph_3P{=}O$$

no trace of 1-methylcyclohexene

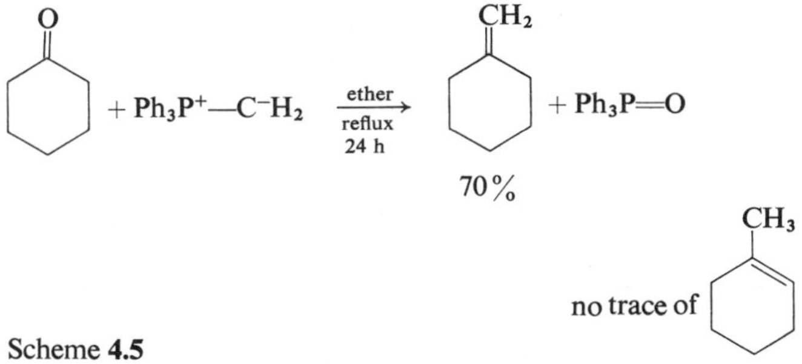

Scheme **4.5**

The formation of olefin and phosphine oxide from ylid and carbonyl compounds is a two-stage process involving an intermediate betaine (4.70). The involvement of the intermediate betaine in the reaction is supported by a considerable amount of evidence, including the isolation of several betaines. A compound which appears to be the betaine (4.72) has been isolated in 70 per cent yield from the reaction of methylenetri(*p*-anisyl)phosphorane (4.71) with benzaldehyde. It is thought that the electron-donating *p*-methoxyl groups so reduce the positive charge on the phosphorus atom that nucleophilic attack by oxygen in (4.72) is stopped, permitting isolation of the betaine itself.

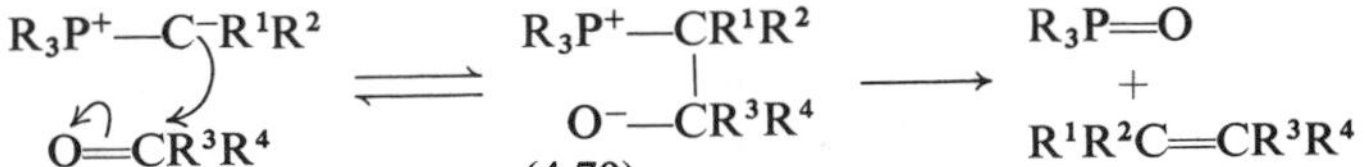

$$R_3P^+\text{—}C^-R^1R^2 + O{=}CR^3R^4 \rightleftharpoons \underset{(4.70)}{R_3P^+\text{—}CR^1R^2\text{—}CR^3R^4\text{—}O^-} \longrightarrow R_3P{=}O + R^1R^2C{=}CR^3R^4$$

$$\underset{(4.71)}{(p\text{-MeOC}_6H_4)_3P^+\text{—}C^-H_2} + C_6H_5CHO \longrightarrow \underset{(4.72)}{(p\text{-MeOC}_6H_4)_3P^+\text{—}CH_2\text{—}CHPh\text{—}O^-}$$

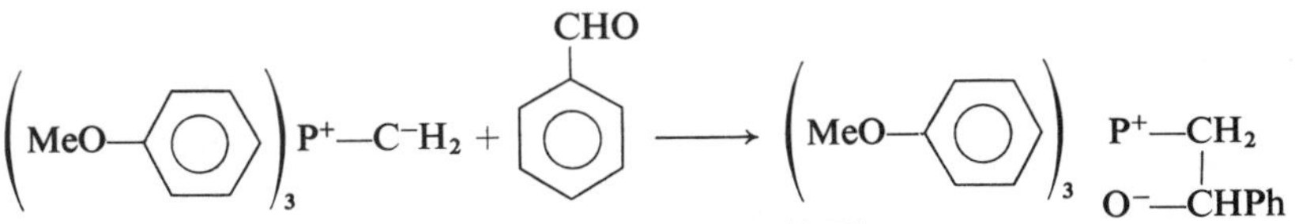

Recently, Matthews and Birum have carefully examined the structure of the supposed betaine intermediate (4.75) isolated from the reaction of hexafluoroacetone (4.74) with the bis-ylid (4.73). While the earlier examples of isolated betaines (e.g. 4.72) almost certainly have acyclic ionic structures, (4.75) appears to have a cyclic covalent one. Phosphorus-31 nuclear magnetic resonance showed that it possessed two different phosphorus atoms, one of which showed a chemical shift position ($\delta^{31}P + 54$ p.p.m.) close to that found for pentacovalent-phosphorus compounds; this obviously supports structure (4.75) rather than the alternative symmetrical structure (4.76).

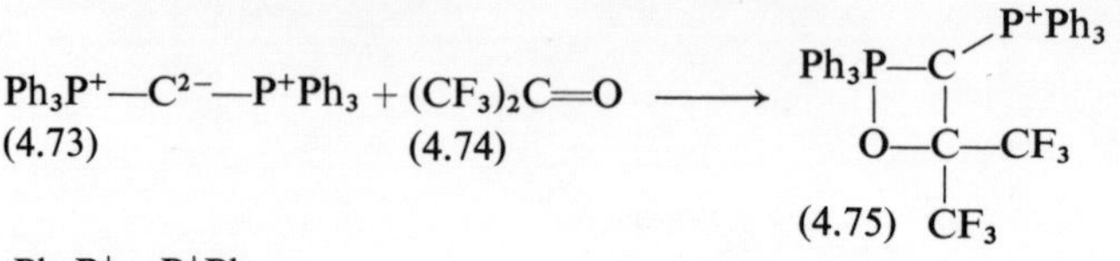

$$\underset{(4.73)}{Ph_3P^+{-}C^{2-}{-}P^+Ph_3} + \underset{(4.74)}{(CF_3)_2C{=}O} \longrightarrow \underset{(4.75)}{\begin{array}{l} \quad\quad\quad\;\; P^+Ph_3 \\ Ph_3P{-}C \\ \;\;\;\;\;\;|\quad\; | \\ \;\;\;\;\;\;O{-}C{-}CF_3 \\ \quad\quad\quad\; | \\ \quad\quad\quad CF_3 \end{array}}$$

$$\underset{(4.76)}{\begin{array}{c} Ph_3P^+ \quad P^+Ph_3 \\ \diagdown\;\diagup \\ C^- \\ | \\ CF_3{-}C{-}CF_3 \\ | \\ O^- \end{array}}$$

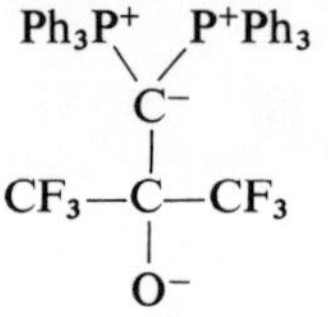

The second stage of the reaction, decomposition of the betaine (4.70) to phosphine oxide and olefin, is thought to proceed through a four-membered cyclic intermediate like (4.75), which is formed by attack of the oxy-anion on the phosphonium group. The driving force for this second state is, presumably, the formation of a strong P=O bond (see p. 33).

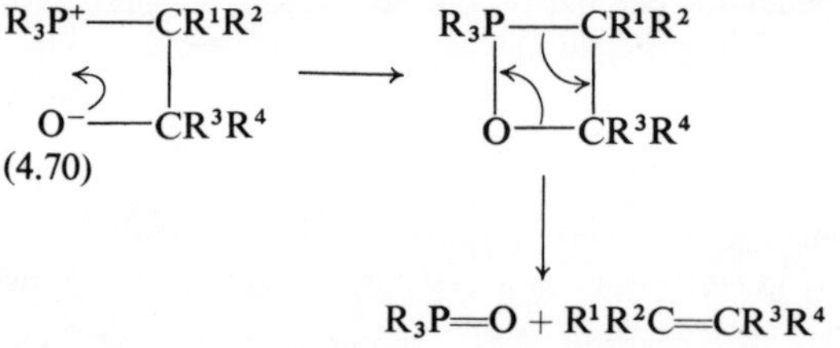

$$\underset{(4.70)}{\begin{array}{ll} R_3P^+ & {-}CR^1R^2 \\ & \;\;| \\ O^- & {-}CR^3R^4 \end{array}} \longrightarrow \begin{array}{ll} R_3P & {-}CR^1R^2 \\ | & \;\;| \\ O & {-}CR^3R^4 \end{array}$$

$$\downarrow$$

$$R_3P{=}O + R^1R^2C{=}CR^3R^4$$

(a) *Stereochemistry of the Wittig reaction.* When the Wittig reaction is used synthetically, it is often important to know whether the olefins produced exhibit *cis* or *trans* stereochemistry. Although this can obviously be determined after the reaction, a stereospecific synthesis (one which produces either a pure *cis*-olefin or a pure *trans*-olefin) would frequently be invaluable, negating the need for lengthy separation procedures. This explains the popularity and importance of all work aimed at understanding and controlling the stereochemistry of the Wittig reaction.

A phosphonium ylid and an aldehyde (or ketone) can react to give two diastereomeric betaines (4.77 and 4.78). The relative rates of formation and

$$R_3P^+\text{—}C^-HR^1 + O\text{=}CHR^2 \rightleftharpoons \text{betaine (4.77)} \longrightarrow cis\text{-}R^1CH\text{=}CHR^2$$

$$R_3P^+\text{—}C^-HR^1 + O\text{=}CHR^2 \rightleftharpoons \text{betaine (4.78)} \longrightarrow trans\text{-}R^1CH\text{=}CHR^2$$

(4.77) (4.78)

decomposition of these betaines control the stereochemistry of the olefin mixture produced by the Wittig reaction; decomposition of betaine (4.77) will lead to *cis*-olefin, while decomposition of betaine (4.78) will lead to *trans*-olefin.

In the absence of solvating effects, the conformations of the betaines are determined by the electrostatic attraction between the negatively charged oxygen atom and the positively charged phosphorus atom. In this eclipsed conformation, betaine (4.78) will obviously be favoured *thermodynamically* because it has a minimum of steric interactions. However, unless the steps in betaine formation are reversible, thermodynamic stability will not matter, and the proportion of each diastereomeric betaine present will depend on their relative rates of formation (i.e. the formation will be *kinetically* controlled). Because of this, it is important to determine whether betaine formation is reversible.

$$Bu_3P + Ph\text{—}\overset{O}{CH\text{—}CH}\text{—}COOEt \longrightarrow Bu_3P^+\text{—}CH(COOEt)\text{—}CH(O^-)Ph \longrightarrow Bu_3P\text{=}O + PhCH\text{=}CH.COOEt$$

(4.79) (4.80) (4.81) (4.82) (4.83)

$$(4.81) \rightleftharpoons PhCHO + Bu_3P^+\text{—}C^-H\text{—}COOEt \xrightarrow{m\text{-}ClC_6H_4CHO} m\text{-}ClC_6H_4CH\text{=}CH\text{—}COOEt + Bu_3P\text{=}O$$

(4.84) (4.85)

(i) Stabilized ylids. Speziale and Bissing reacted tributylphosphine (4.79) with ethyl *trans*-phenylglycidate (4.80) in the presence of *m*-chlorobenzaldehyde to show that stabilized phosphonium ylids (see p. 139) undergo reversible betaine formation. The epoxide undergoes nucleophilic attack by the phosphine to give the betaine (4.81), which then either decomposes to phosphine oxide (4.82) and olefin (4.83), or dissociates to the phosphonium ylid (4.84) and benzaldehyde. *m*-Chlorobenzaldehyde is very much more reactive than benzaldehyde to nucleophilic attack, so any ylid (4.84) formed will react with the former rather than the latter. The amount of *m*-chlorophenyl-substituted olefin (4.85) formed in this way will be a measure of the betaine dissociation.

Stabilized phosphonium ylids (e.g. 4.84) produce mostly the *trans*-olefin when reacted with an aldehyde (or ketone). This is predictable if the reaction pathway is controlled by the thermodynamic stability of the intermediate betaine, the betaine-formation step being both reversible and rate determining (i.e. slower than betaine decomposition).

Although the effect of different solvents on the *cis*/*trans* ratio has not been fully resolved, it appears that protonic solvents like methanol increase the proportion of *cis*-olefin (see Table 25). It is suggested that solvents which solvate the oxy-anion in the intermediate betaine (4.86), reduce the electrostatic attraction between O^- and P^+; this allows other conformations, in which the other diastereomeric betaine will be more stable, to become important.

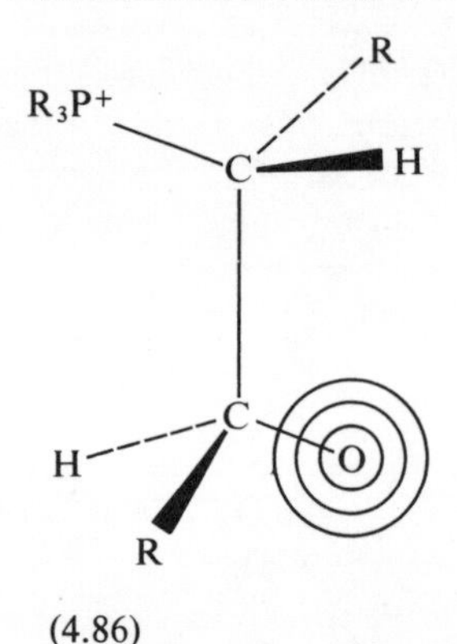

(4.86)

(ii) Reactive and partially stabilized ylids. Reactive ylids and partially stabilized ylids (see p. 140) both undergo the Wittig reaction to give mixtures of *cis*- and *trans*-olefins, the proportions of which are largely dependent on the reaction conditions.

Table 25 Effect of solvent on *cis/trans* ratios from stable ylids

Reaction:

$Ph_3P^+—CH^-—COOMe + CH_3CHO \rightarrow Ph_3P{=}O + CH_3CH{=}CHCOOMe$

Solvent	*Overall yield*	% cis	% trans
CH_2Cl_2	88%	6	94
DMF	98%	3	97
MeOH	96%	38	62

By preparing one of the diastereomeric betaines free of the other, Trippett has shown that, for these ylids, betaine formation is reversible. He reacted *trans*-stilbene oxide (4.87) with sodium diphenylphosphide, which on acidification gave the β-hydroxyphosphine (4.88). Treatment of this with methyl iodide gave the hydroxyphosphonium salt (4.89). The addition of one mole of base to this salt gave the betaine (4.90), the stereochemistry of which could only give *cis*-stilbene, unless it dissociated back to the ylid and aldehyde. These could then react to produce the other diastereomeric betaine (4.91), which in turn would give *trans*-olefin. Trippett reasoned that the formation of any *trans*-olefin would prove that betaine formation was reversible for these ylids (he had already shown that the betaines (4.90 and 4.91) were not directly interconvertible without the intermediate formation of ylid and aldehyde). The stilbene mixture that he obtained consisted of approximately equal quantities of *cis*- and *trans*-olefin, proving that betaine formation is reversible for reactive ylids.

$Ph_2P^-Na^+$ + (4.87) ⟶ [betaine alkoxide] —AcOH→ (4.88) —MeI→ (4.89)

Ph H C O C Ph H (4.87)

Ph_2P H C Ph C O^- H Ph

AcOH

Ph_2P H C Ph C HO H Ph (4.88)

MeI

Me Ph_2P^+ H C Ph I^- C HO H Ph (4.89)

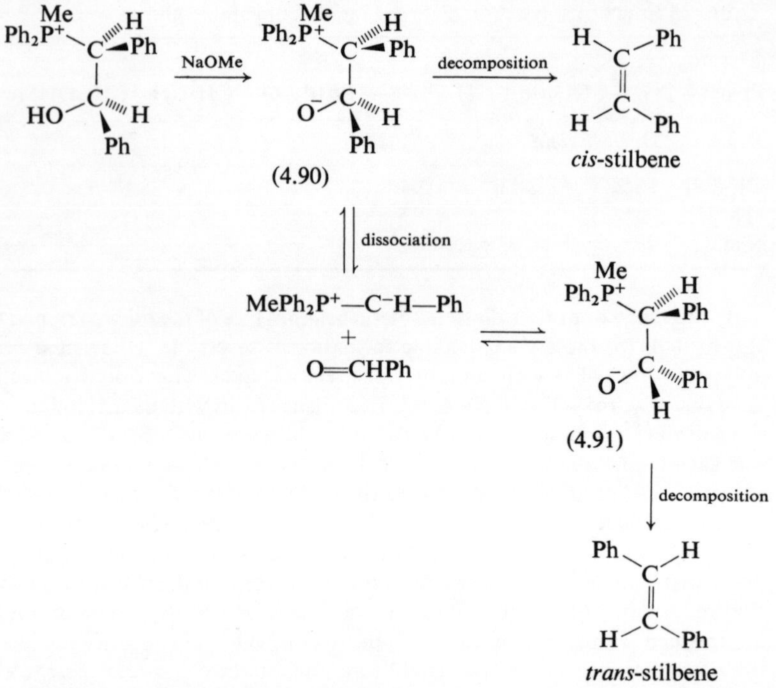

$Ph_2P^+(Me)$–CH(Ph)–CH(Ph)–OH $\xrightarrow{\text{NaOMe}}$ $Ph_2P^+(Me)$–CH(Ph)–CH(Ph)–O^- (4.90) $\xrightarrow{\text{decomposition}}$ HPhC=CHPh

cis-stilbene

(4.90) $\rightleftharpoons$ (dissociation)

$MePh_2P^+$—C^-H—Ph + O=CHPh $\rightleftharpoons$ $Ph_2P^+(Me)$–CH(Ph)–CH(Ph)–O^- (4.91)

(4.91) $\xrightarrow{\text{decomposition}}$ PhHC=CHPh

trans-stilbene

Recently, Russian workers have shown that the rate of olefin formation increases with increase in electrophilicity of the aldehyde used. They interpreted these results as meaning that betaine formation, not betaine decomposition, was rate determining, since increasing electrophilicity of the aldehyde would obviously facilitate the attack by an ylid carbanion.

(iii) Salt and solvent effects. The effect of reaction conditions on the *cis/trans* ratios in the Wittig reaction has been the subject of much, apparently contradictory, work during the past eight years. Sometimes different workers obtained different results from apparently identical experiments. Recently, much of this confusion has been dispersed by realization that alkali-metal salts can have fundamental effects on the stereochemistry of the reaction.

Reactive ylids are normally generated from their phosphonium salts by treatment with a metal alkyl, usually a lithium alkyl (because of the relative ease of preparation). Therefore studies of such ylids have invariably been carried out in the presence of lithium salts, and the results of many early experiments should be reinterpreted in the light of recent work with solutions of ylids from which metal salts have been vigorously excluded.

Reactive ylids taking part in Wittig reactions in polar, aprotic solvents like dimethylformamide are little affected by lithium salts and give largely *cis*-olefins. However, in non-polar solvents the same reactions show a considerable salt dependence (see Table 26).

Table 26 Salt effects on Wittig reactions in non-polar solvents

Reaction: $Ph_3P^+—C^-H—CH_3 + PhCHO \rightarrow Ph_3P{=}O + CH_3CH{=}CH—Ph$

Salt present	*Overall yield of olefin*	cis/trans *ratio*
salt free	98%	87/13
LiCl	70%	81/19
LiBr	68%	61/39
LiI	76%	58/42
$Li(Ph_4B)$	63%	50/50

Reaction:
$Ph_3P^+—C^-H—CH_2CH_3 + PhCHO \rightarrow Ph_3P{=}O + CH_3CH_2CH{=}CHPh$

Salt present	*Overall yield of olefin*	cis/trans *ratio*
salt free	88%	96/4
LiCl	80%	90/10
LiBr	80%	86/14
LiI	81%	83/17
$Li(Ph_4B)$	60%	52/48

The reactions were carried out in benzene–light petroleum (4/1) at 0 °C.

In the absence of lithium salts, Wittig reactions with reactive ylids in non-polar solvents again give largely *cis*-olefins; this reaction has been used for the synthesis of a number of naturally occurring compounds containing a *cis* double bond.

In non-polar solvents, the addition of lithium salts increases the proportion of *trans*-isomer in the overall yield of olefin produced from a Wittig reaction involving reactive ylids (see Table 26). The increase in the proportion of *trans*-

$Ph_3P^+—C^-H—R$ + PhCHO (slow ⇌)

(4.92) fast → cis-olefin

(4.93) fast → trans-olefin

olefin is also dependent on the size of the anion associated with the lithium ion. Under salt-free conditions the first stage of the Wittig reaction (i.e. betaine formation) is thought to be rate determining and, since the betaine decomposes rapidly to an olefin and phosphine oxide, the *cis/trans* ratio will be controlled by the rates of formation of the two diastereomeric betaines (4.92 and 4.93). The rate of formation of betaine (4.92) is faster than that of betaine (4.93), so a high proportion of *cis*-olefin is formed.

In the presence of lithium salts the betaines (4.92 and 4.93) are largely coordinated with lithium ions, and their rate of decomposition is reduced to the extent that it becomes comparable with their rate of formation. At this point betaine reversibility becomes important, and the proportions of betaines (4.92 and 4.93) present depend on their thermodynamic stability rather than their rate of formation. Betaine (4.93), which leads to *trans*-olefin, is thermodynamically more stable and so the proportion of *trans*-olefin increases.

The dependence of olefin stereochemistry on time has lent support to this explanation. Table 27 shows the variation of *cis/trans* ratio with time for the reaction of *n*-propylidenetriphenylphosphorane with benzaldehyde. Because of its greater rate of formation, betaine (4.92) is initially in high concentration and, even in the presence of lithium salts, the olefin obtained is largely *cis*. However, after an hour the olefin is largely *trans* because the betaines (4.92 and 4.93) have had time to reach equilibrium.

Table 27 Variation of *cis/trans* ratio with time

Reaction: Ph_3P^+—CH—Et + PhCHO → Ph_3PO + PhCH=CHEt

Time	*Salt-free conditions*	*With added* LiI
	cis/trans	cis/trans
1 min	90/10	68/32
1 h	91/9	35/65
15 h		33/67

4.2.2 *Other reactions of phosphonium ylids*

(a) *Hydrolysis and alcoholysis.* Phosphonium ylids (4.94) undergo hydrolysis to a hydrocarbon and phosphine oxide in the same way as phosphonium salts, which are almost certainly formed as intermediates. At least a pre-equilibrium with the parent phosphonium salt is suggested by the decomposition of *p*-nitrobenzylidenetriphenylphosphorane (4.95) in *O*-deuterated ethanol, catalysed by sodium ethoxide; this decomposition gives both dideutero- and trideutero-*p*-nitrotoluene as products.

$R_3P^+—C^-H—R' + H_2O \rightleftharpoons R_3P^+—CH_2R'$
(4.94)

^-OH

$\downarrow$

$[R_3P—CH_2—R' \quad O—H]$ ^-OH

$R_3PO + R'—C^-H_2 + H_2O$

$\downarrow H_2O$

$R'CH_3 + OH^-$

$Ph_3P^+—C^-H—C_6H_4—NO_2$

(4.95)

$EtO^- \rightleftharpoons EtOD$

EtO^-
$Ph_3P^+—CHD—C_6H_4—NO_2 \longrightarrow Ph_3P{=}O + CHD_2—C_6H_4—NO_2$

$\rightleftharpoons$

$Ph_3P^+—C^-D—C_6H_4—NO_2$

$EtO^- \rightleftharpoons EtOD$

$Ph_3P^+—CD_2—C_6H_4—NO_2 \longrightarrow Ph_3P{=}O + CD_3—C_6H_4—NO_2$

The ease of hydrolysis of the ylid depends on the nature of the substituent R′; when this is alkyl (e.g. 4.96), violent reaction takes place with cold water, while electron-withdrawing substituents (e.g. 4.97) may make the ylid stable to hot alkali.

$$Ph_3P^+—C^-H—CH_3$$
(4.96)

$$Ph_3P^+—C^-(CO—CH_3)_2$$
(4.97)

Hydrolysis reactions of phosphonium ylids and phosphonium salts are important, since a number of useful synthetic methods involve these hydrolyses as a final step (see p. 156).

(b) *Oxidation.* The most reactive phosphonium ylids must be prepared and reacted under a nitrogen atmosphere because they react rapidly with oxygen to give, initially, phosphine oxide (4.98) and an aldehyde (or ketone). Bestmann showed that, in a limited supply of oxygen, the products also included the olefin (4.99); he developed this into a useful method of olefin synthesis. The reaction presumably takes place by attack of the ylid on an electrophilic oxygen atom, followed by a four-centred elimination. In a limited supply of oxygen, the carbonyl compound (4.100) thus formed can undergo a Wittig reaction with an excess of ylid to give an olefin.

$$R_3P^+—C^-HR' + O_2 \longrightarrow R_3P{=}O + R'—CHO$$
(4.98)

$$2R_3P^+—C^-H—R' \xrightarrow{O_2} R'—CH{=}CH—R' + 2R_3P{=}O$$
(4.99)

$$R_3P^+—C^-H—R' + O_2 \longrightarrow R_3P^+—CH(R')—O—O^-$$

$$\downarrow$$

$$R_3P{=}O + R'CHO$$
(4.100)

$$\downarrow R_3P^+—C^-HR'$$

$$R_3P{=}O + R'CH{=}CHR'$$

Stable ylids will not react with oxygen (therefore they can be exposed to air without decomposition), although they will react with ozone, even at –70 °C. The ylid (4.101) gives an α-diketone (4.102), phosphine oxide and oxygen, possibly by the mechanism shown (Scheme **4.6**).

$$Ph_3P^+—\underset{\underset{R}{|}}{C^-}—COPh \xrightarrow[CH_2Cl_2]{O_3} \begin{array}{c} Ph_3P—\overset{\overset{R}{|}}{C}—CO—Ph \\ | \quad\quad | \\ O^-—O^+—O \end{array}$$

(4.101)

$$\downarrow$$

$$[Ph_3P{=}O^+—O^-] + RCOCOPh$$

(4.102)

$$\downarrow$$

Scheme **4.6**

$$Ph_3P{=}O + \tfrac{1}{2}O_2$$

Treatment of stable ylids with peracids gives olefins and phosphine oxide; this can be used as a method of synthesis complementary to the reaction with oxygen, which will only work with reactive ylids. Unfortunately, the yields from this reaction are variable.

$$2Ph_3P^+—\underset{\underset{R}{|}}{C^-}—CO—R' + CH_3COOOH$$

$$\downarrow$$

$$2Ph_3P{=}O + R'—CO—\underset{\underset{R}{|}}{C}{=}\underset{\underset{R}{|}}{C}—CO—R'$$

(c) *Alkylation and acylation.* Alkylation of ylids with alkyl halides normally takes place at the α-carbon atom and is an important route to more complex ylids (4.103) and phosphonium salts (4.104), which are often difficult or impossible to obtain by direct quaternization. Alkylidene ylids (4.105; R^1 = alkyl) give the expected phosphonium salts (4.104; R^1 = alkyl), and intramolecular reactions have been used to prepare cyclic compounds (e.g. 4.106).

$$Ph_3P^+—C^-H—R^1 + R^2X \longrightarrow Ph_3P^+—\underset{\underset{R^2}{|}}{CH}—R^1 \quad X^-$$

(4.105) (4.104)

$$\downarrow \text{base}$$

$$Ph_3P^+—\underset{\underset{R^2}{|}}{C^-}—R^1$$

(4.103)

$$Ph_3P^+{-}C^-H{-}(CH_2)_4CH_2Br \longrightarrow \underset{Br^-}{Ph_3P^+{-}CH}\langle\begin{matrix}CH_2{-}CH_2\\CH_2{-}CH_2\end{matrix}\rangle CH_2$$

(4.106)

Complications arise with stabilized ylids of the type (4.107), since these are resonance hybrids with canonical forms (4.107a and b) and so have negative charge density at both the carbon and the oxygen atoms. Ester ylids (4.107; $R' = OR^2$) alkylate normally at a carbon atom to give (4.108) and probably have very little contribution from (4.107b). β-Keto ylids (4.107; R′ = alkyl or aryl), on the other hand, generally give the oxygen-alkylated product (4.109).

$$R_3P^+{-}C^-H{-}\underset{\displaystyle O}{\overset{}{\underset{\Vert}{C}}}{-}R' \longleftrightarrow R_3P^+{-}CH{=}\underset{\displaystyle O^-}{\underset{|}{C}}{-}R'$$

(4.107a) (4.107b)

$$R_3P^+{-}C^-H{-}\underset{\displaystyle O}{\underset{\Vert}{C}}{-}O{-}R^2 \xrightarrow{R^3X} \underset{X^-}{R_3P^+}{-}\underset{\displaystyle R^3}{\underset{|}{CH}}{-}\underset{\displaystyle O}{\underset{\Vert}{C}}{-}OR^2$$

(4.108)

$$R_3P^+{-}C^-H{-}\underset{\displaystyle O}{\underset{\Vert}{C}}{-}R^1$$

$$\updownarrow$$

$$R_3P^+{-}CH{=}\underset{\displaystyle O^-}{\underset{|}{C}}{-}R^1 \xrightarrow{R^2X} \underset{X^-}{R_3P^+}{-}CH{=}\underset{\displaystyle OR^2}{\underset{|}{C}}{-}R^1$$

(4.109)

$$\underset{(4.110)}{\underset{Br^-}{Ph_3P^+{-}CH_3}} \xrightarrow[\text{ether}]{\text{LiBu}} \underset{(4.111)}{Ph_3P^+{-}C^-H_2}$$

$$\downarrow PhCOCl$$

$$\underset{(4.112)}{Ph_3P^+{-}CH_2{-}CO{-}Ph\ \ Cl^-}$$

$$\downarrow Ph_3P^+{-}C^-H_2$$

$$\underset{(4.113)}{Ph_3P^+{-}C^-H{-}COPh} + Ph_3P^+{-}CH_3\ \ Cl^-$$

The acylation of phosphonium ylids proceeds in a way similar to alkylation, with the important difference that any α-hydrogens in the initially formed salt will be more acidic than those in the parent phosphonium salt of the original ylid. For example, methylenetriphenylphosphorane (4.111) reacts with benzoyl chloride to give initially the phosphonium salt (4.112). However, this has an α-hydrogen which is more acidic than those in the salt (4.110), and so reacts with the ylid (4.111) to form a new ylid (4.113). Therefore the product isolated from the reaction of the ylid (4.111) with half a mole of benzoyl chloride will be the ylid (4.113), and not its phosphonium salt (4.112). This is a completely general reaction involving the equilibration of ylids according to the acidity of their phosphonium salts; the reaction is known as *transylidation*.

$R_3P^+—C^-H—CO—Ph$
(4.114)

Acylations involving benzoyl halides have been widely used to prepare β-keto ylids (4.114). However, the reaction suffers from the disadvantage that since half of the original ylid is required to convert the acyl phosphonium salt to its ylid, the yield can never be greater than 50 per cent. Since the phosphonium ylid is the most expensive and the least readily available of the reactants, attempts have been made to find alternative acylating agents. Carboxylic acid esters readily acylate reactive ylids, initially in the same way as acid halides; but the β-keto salt (4.115) so formed is associated with an ethoxide ion, which removes a proton to give the β-keto ylid (4.116). Because the anion originally displaced can act as a base, transylidation does not take place, and the theoretical yield of ylid (4.116) is 100 per cent.

$$Ph_3P^+—C^-H_2 + R—COOEt \longrightarrow \underset{(4.115)}{Ph_3P^+—CH_2—CO—R \quad OEt^-}$$

$$\downarrow$$

$$\underset{(4.116)}{Ph_3P^+—C^-H—CO—R} + EtOH$$

Formate esters will carry out formylation in a similar way, but only when the ylid is added to an excess of formate ester (reverse addition). If the more usual method of addition of the ester to an excess of ylid is used, the carbonyl group of the formate ester undergoes a Wittig reaction to give a vinyl ether (4.117). When formate is in excess, the betaine (4.118) is probably solvated to such an extent that it cannot undergo a Wittig reaction, and the alternative pathway, to give the formylated product, predominates.

$$Ph_3P^+{-}C^-H{-}R + H{-}CO{-}OEt \longrightarrow Ph_3P^+{-}CH(R){-}CO{-}H \quad OEt^-$$

$$\downarrow$$

$$Ph_3P^+{-}C^-R{-}CO{-}H + EtOH$$

$$Ph_3P^+{-}\overset{-}{C}_6H_{10} + HCOOEt \longrightarrow Ph_3P^+{-}C_6H_{10}{-}C(O^-)(H)(OEt)$$

(4.118)

$$\downarrow$$

$$Ph_3P{=}O + EtOCH{=}C_6H_{10}$$

(4.117)

As with alkylation, β-keto ylids can undergo acylation at either the oxygen atom or a carbon atom. Acid anhydrides generally acylate at a carbon atom, while acid chlorides frequently attack the oxygen atom.

$$Ph_3P^+{-}C^-H{-}COR \xrightarrow{(R'{-}CO)_2O} Ph_3P^+{-}\bar{C}(COR)(COR')$$

$$\downarrow R'COCl$$

$$Ph_3P^+{-}\overset{H}{C}{=}C(R){-}OCOR' \quad Cl^-$$

Alkylation and acylation reactions have been extensively studied because, as well as providing routes to a variety of ylids and phosphonium salts, they can be used to synthesize a wide range of organic compounds *not* containing phosphorus. Alkaline hydrolysis (see p. 129) of the phosphonium salts prepared by alkylation or acylation reactions removes the phosphorus atom to give a suitably substituted organic side chain (4.119). Table 28 outlines a variety of alkylation and acylation reactions.

$$R_3P^+{-}CH(R^1)(R^2) \xrightarrow[\text{aqueous NaOH}]{\text{reflux}} R_3P{=}O + CH_2(R^1)(R^2)$$

(4.119)

Table 28 Alkylation and acylation of phosphonium ylids

1. $Ph_3P^+—C^-H_2$ + (cyclopropane-1,2-diyl)bis(CH_2OTs) $\longrightarrow$ $Ph_3P^+—CH$(bicyclic) OTs^-

2. 1-(CH_2Br)-8-($CH_2P^+Ph_3$ Br^-)naphthalene $\xrightarrow{NaOEt}$ (P^+Ph_3 Br^-)acenaphthene $\xrightarrow{heat}$ acenaphthylene 66%

3. $Ph_3P^+—C^-H—Ph + Et_3O^+BF_4^-$ $\longrightarrow$ $Ph_3P^+—CH(Et)—Ph\ \ BF_4^-$

4. $Ph_3P^+—C^-H—R' + RCOSEt$ $\longrightarrow$ $Ph_3P^+—C^-(R')—COR + EtSH$

5. $Ph_3P^+—CH_2(CH_2)_nCH_2COOEt\ Br^-$ $\xrightarrow{base}$ $Ph_3P^+—C^-H(CH_2)_nCH_2COOEt$ $\longrightarrow$ $Ph_3P^+—CH$(cyclic: $(CH_2)_n$, CH_2, C=O) OEt^- $\longrightarrow$ $Ph_3P^+—C^-$(cyclic: $(CH_2)_n$, CH_2, C=O)

6. $Ph_3P^+—C^-(R)—COOEt + R'CH_2COX$ $\longrightarrow$ $Ph_3P{=}O$ + $R(COOEt)C{=}C{=}CHR'$

(d) *Reaction with multiple bonds.* Phosphonium ylids will only react with multiple bonds which are activated by strongly electron-withdrawing substituents (e.g. —CO—R, —COOR), or are between atoms of elements of differing electronegativity (e.g. >C=O, >C=N—).

$$R_3P^+—C^-H—R' + X{=}Y \longrightarrow R_3P^+—CH(R')—X—Y^- \quad (4.120)$$

Initially, the ylid carbanion attacks at the positive end of the double bond to give the 1,4 dipole (4.120) (but see the Wittig reaction p. 142). This dipole can then undergo a variety of reactions, depending on the substituents present; these reactions are illustrated for the carbon–carbon double bond in Scheme **4.7**. If an α-hydrogen atom is available, the most common reaction is a 1,3 proton shift to give the ylid (4.121). However if R′ is a good leaving group (e.g. halogen), elimination readily takes place, giving the allyl salt (4.122) or its ylid. The third pathway, involving the elimination of phosphine, is much less common.

$$Ph_3P^+—C^-HR + R^1CH{=}CHR^2$$

↓

$$[Ph_3P^+—CHR—R^1CH—C^-HR^2]$$

elimination of phosphine → Ph_3P + cyclopropane (CHR, CHR^1, CHR^2)

1,3 proton shift → $Ph_3P^+—C^-R—CHR^1—CH_2R^2$ (4.121)

loss of R^{1-} → $Ph_3P^+—CHR—CH{=}CHR^2$ R^{1-} (4.122)

Scheme **4.7**

In reactions with activated triple bonds, a fourth pathway is available; this involves a rearrangement via the four-centred transition state (4.123) to give the ylid (4.124).

$$Ph_3P^+—C^-H—CO—Ph + EtOOC—C{\equiv}C—COOEt$$

↓

$Ph_3P^+—CH—CO—Ph$ over $C^-{=}C$ with EtOOC and COOEt substituents (four-centred transition state)

(4.123)

↓

$$Ph_3P^+—C^-(COOEt)—C(COOEt){=}CH—CO—Ph$$

(4.124)

(e) *Reduction.* The reactions of ylids and the corresponding phosphonium salts with lithium tetrahydroaluminate lead to different products. This indicates that these reactions, unlike hydrolysis reactions (see p. 150), do not take place via a common intermediate.

$$\underset{\substack{Br^- \\ (4.125)}}{Ph_3P^+—CH_2—Ph} \xrightarrow{LiAlH_4} Ph_3P + CH_3—Ph$$

Phosphonium salts (e.g. 4.125) react to give phosphines, losing that group most stable as an anion. The reaction presumably takes place via an attack by a hydride ion on the positively charged phosphorus atom. Phosphonium ylids are also reduced to phosphines, but they do not necessarily lose the group most stable as an anion. For example, the ylids (4.126) always lose a phenyl group, irrespective of the other substituents on the phosphorus atom. The mechanism of ylid reduction is not understood, but may involve direct displacement by a hydride ion, as in the case of phosphonium salts. The group most stable as an anion (Ph) will still be lost because, unless protonation can first take place, the loss of $R'—CH^{2-}$ would be most unfavourable.

$$Ph_3P^+—C^-H—R' \xrightarrow{LiAlH_4} Ph_2P—CH_2R + PhH$$
(4.126; R'=H, $COCH_3$, COPh, $COC(CH_3)_3$, $COCH(CH_3)_2$)

Stable ylids can be reduced effectively by zinc in acetic acid or by hydrogenation using Raney nickel, and the group lost is that most stable as an anion.

$$Ph_3P^+—C^-H—COR \xrightarrow{Zn/AcOH} Ph_3P + CH_3COR$$

(f) *Other reactions.* Phosphonium ylids react readily with epoxides to give a variety of products depending on the particular ylid and epoxide used. For example, styrene oxide reacts with the ylid (4.127) to give triphenylphosphine oxide and the cyclopropane (4.128). However, cyclohexene oxide reacts under much milder conditions to give only 3 per cent of the equivalent cyclopropane (4.129) and 92 per cent of the olefin (4.130).

PhCH—CH₂ (epoxide, bridged by O) + $Ph_3P^+—C^-H—COOEt$ (4.127)

$$\downarrow 200\,°C$$

$Ph_3P{=}O$ + PhCH—CH—COOEt (cyclopropane, bridged by CH_2)
(4.128)

$n\text{-}Bu_3P^+—C^-H—COOEt$ + (cyclohexene oxide)

benzene, 80 °C

(bicyclo[4.1.0]heptane)—COOEt + (cyclopentyl)—CH═CH—COOEt

(4.129) (4.130)

The mechanisms of these reactions have been extensively studied recently and, in both cases, the ylid carbanion attacks the epoxide ring to give the phosphorus(V) intermediates (4.131 and 4.133). The product derived from styrene oxide then eliminates phosphine oxide to give the cyclopropane via the zwitterion (4.132), while the main pathway from (4.133) involves the rearrangement (4.134).

$Ph_3P^+—C^-H—COOEt + PhCH—CH_2$ (epoxide, O) ⟶ (4.131)

COOEt, Ph_3P, Ph, O

(4.131)

↓

$Ph_3P{=}O$ + CH(COOEt)—CHPh—CH_2 (cyclopropane) ⟵ Ph_3P^+—O—CH_2—CH(Ph)—C^-H—COOEt

(4.132)

$n\text{-}Bu_3P^+—C^-H—COOEt$ + (cyclohexene oxide) ⟶ $n\text{-}Bu_3P$ (CH—COOEt, O, cyclohexane ring)

(4.133)

↓

$n\text{-}Bu_3P{=}O$ + (cyclopentyl)—CH═CH—COOEt ⟵ $n\text{-}Bu_3P^+$—O—(cyclohexyl)—C^-HCOOEt

(4.134)

Certain phosphonium ylids decompose spontaneously at room temperature. α-Alkoxyphosphonium salts with an alkoxy group of butoxy (4.135; R = Pr) or larger, react with strong bases to form ylids, but these ylids rapidly decompose to give the products shown in Scheme **4.8**.

$$Ph_3P^+{-}CH_2OCH_2R \;\; X^- \quad (4.135)$$

$$\downarrow \text{BuLi, ether } 10\,^\circ C$$

$$[Ph_3P^+{-}C^-H{-}OCH_2R] \quad (4.136)$$

$$\downarrow$$

$$Ph_3P + RCH_2O{-}CH{=}CH{-}OCH_2R + RCH{=}CH{-}OCH_2R + (RCH_2O)_2CH_2 + RCH_2OH$$

Scheme **4.8**

The actual intermediacy of the ylid (4.136) has not been proved and the mechanism is still in some doubt, although one possibility involves decomposition of the ylid to give a carbene (4.137) and triphenylphosphine. The formation of carbenes from phosphonium ylids on irradiation with ultraviolet light is well established and, in this instance, the decomposition will be facilitated by resonance stabilization of the carbene. Phosphonium salts (4.138) which are substituted in the β-position with good anionic leaving groups give ylids which rapidly decompose by elimination to produce vinylphosphonium salts (4.139).

$$[Ph_3P^+{-}C^-H{-}OCH_2R]$$

$$\downarrow$$

$$Ph_3P + :CH{-}\ddot{\underset{..}{O}}{-}CH_2R \longleftrightarrow C^-H{=}O^+{-}CH_2R \quad (4.137)$$

$$\underset{(4.138;\ Y = \text{halide, OMe, } P^+R_3)}{Ph_3P^+{-}CH_2{-}CH_2{-}Y \;\; X^-} \xrightarrow{\text{base}} Ph_3P^+{-}C^-H{-}CH_2{-}Y$$

$$\downarrow$$

$$Ph_3P^+{-}CH{=}CH_2 \;\; Y^- \quad (4.139)$$

In conclusion, we can say that the carbanionic centre of phosphonium ylids will react with almost any electrophile, depending on the degree of stabilization by the substituents on the α-carbon atom of the ylid.

4.3 Imidophosphoranes

Compounds analogous to phosphonium ylids, but with the carbanionic centre replaced by negatively charged nitrogen, are well known. The direct analogues of phosphonium ylids are known as imidophosphoranes (4.140) and presumably exist, like their carbon analogues, as resonance hybrids, with contributing forms (4.140a and b).

$R_3P^+—N^-—R' \longleftrightarrow R_3P=N—R'$

(4.140a) (4.140b)

Since nitrogen possesses two lone pairs of electrons in these compounds, it can be considered in three alternative hybridized states rather than two, as proposed for phosphonium ylids (see p. 136). The nitrogen atom could be sp (4.141), sp^2 (4.142) or sp^3 (4.143) hybridized, although analogies with phosphonium ylids would favour (4.142). Very recently, an X-ray study carried out on *N*-methylfluorodiphenylimidophosphorane (4.144) revealed a P—N—C bond angle of 119° which suggests sp^2-hybridized nitrogen. The length of the P—N bond, $1{\cdot}641 \times 10^{-10}$ m (1·641 Å), is much less than the expected single-bond length, $1{\cdot}78 \times 10^{-10}$ m (1·78 Å), and closer to the double-bond length, $1{\cdot}64 \times 10^{-10}$ m (1·64 Å), which was calculated from covalent radii (cf. phosphonium ylids, p. 141).

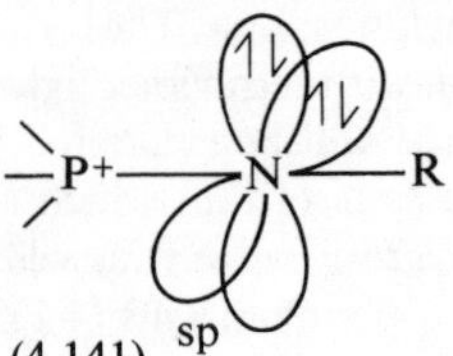

$—P^+—N—R$ sp

(4.141)

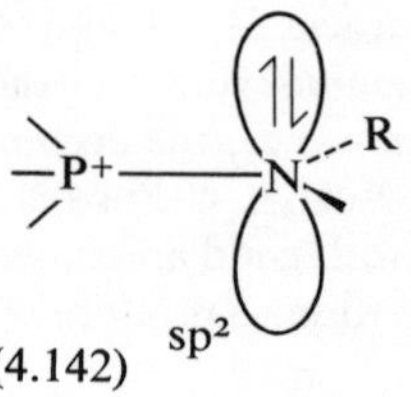

$—P^+—N$ R sp^2

(4.142)

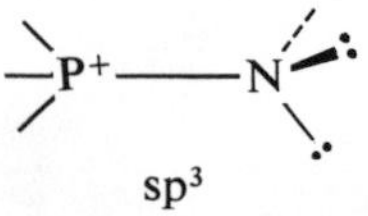

$—P^+—N$ sp^3

(4.143)

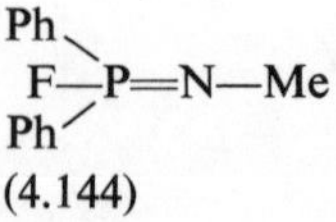

Ph₂FP=N—Me

(4.144)

Ph, F, Ph—P—N 1·641, 119°, Me

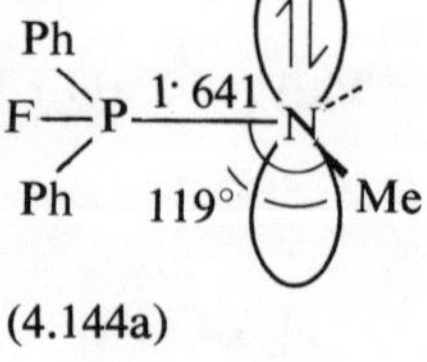

(4.144a)

The dipole moment of $1{\cdot}47 \times 10^{-29}$ C m (4·40 debyes) suggests that the P—N bond is quite highly polarized, but there could still be a large contribution from (4.140b), since the dipole may well derive mainly from σ-electron displacement (cf. bonding in transition-metal complexes). The dissociation energies of the P=N bond are high, 380–550 kJ mol^{-1} (90–130 kcal mol^{-1}), but they are lower than those of the P=O bond, 630 kJ mol^{-1} (150 kcal mol^{-1}) (see pp. 33–4).

$>P^+—\bar{C}<$ (4.145) $>P^+—\bar{N}<$ (4.146) $>P^+—\bar{O}<$ (4.147)

Phosphonium ylids (4.145), imidophosphoranes (4.146) and phosphine oxides (4.147) are all isoelectronic, and might be expected to be chemically related. Without entering into a long discussion on this point, suffice it to say that the analogies are definitely observed, but there is a large reduction in reactivity at both the phosphorus atom and the other heteroatom in going from (4.145) to (4.146) and from (4.146) to (4.147).

A variety of compounds containing the P=N link are known (we have already met one group, the phosphonitriles, Chapter 3, p. 120) and representatives have been tabulated in Table 29, in an attempt to demonstrate the range of these compounds and clarify their nomenclature.

Table 29

$Ph_3P=N—Ph$	*N*-phenyltriphenylimidophosphorane (triphenylphosphine phenyl imine)
$Me_3P=N—COPh$	*N*-benzoyltrimethylimidophosphorane (trimethylphosphine benzoyl imine)
$Ph_3P=N—SO_2—C_6H_4—CH_3$	toluene-*p*-sulphonyl-triphenylimidophosphorane
$Ph_3P=N—N=CPh_2$	triphenylphosphine benzophenonazine

4.3.1 *Formation*

Imidophosphoranes can be prepared by methods analogous to those used for phosphonium salts but specific methods are also available. These, or analogous preparations, have been discussed in principle in other chapters, and a summary is given in Table 30.

Table 30 Preparation of imidophosphoranes

1. $$Ph_3P^+{-}NH_2\ HSO_4^- \xrightarrow[\text{liquid } NH_3]{NaNH_2} Ph_3P{=}NH \quad (92\%\ \text{m.p. } 126\,°C)$$

$$Ph_3P + H_2N{-}SO_3H \longrightarrow Ph_3P^+{-}NH_2\ HSO_4^- \xrightarrow[\text{liquid } NH_3]{NaNH_2} Ph_3P{=}NH$$

$$Ph_3P + PhCOONHCOPh \xrightarrow[100\,°C]{24\text{ h}} Ph_3P{=}N{-}CO{-}Ph + PhCOOH$$

$$R_3P{:} + Cl{-}N(H){-}SO_2{-}Ar \longrightarrow R_3P{=}NSO_2Ar$$

(e.g. chloramine-T
Ar = *p*-tolyl)

2. $$Ph_3PBr_2 + PhNH_2 \xrightarrow[\text{reflux 20 min}]{2Et_3N\ CCl_4} Ph_3P{=}N{-}Ph + 2Et_3NH^+Br^- \quad (80\%\ \text{m.p. } 131\,°C)$$

This is probably the most generally applicable method.

$$Ph_3PBr_2 + Ph_3P{=}NH \longrightarrow Ph_3P{=}N{-}P^+Ph_3\ \ Br^-$$

$$Ph_3PBr_2 + NH_2.NH_2 \xrightarrow{NaNH_2} Ph_3P{=}N{-}NH_2 \xrightarrow{\text{excess } Ph_3PBr_2} Ph_3P{=}N{-}N{=}PPh_3$$

$$Ph_3PCl_2 + H_2N{-}N{=}C\langle^{R^1}_{R^2} \xrightarrow[-10\,°C]{Et_3N} Ph_3P{=}N{-}N{=}C\langle^{R^1}_{R^2}$$

$$PCl_5 + Cl_2P(=O){-}NHR \longrightarrow Cl_3P{=}N{-}R + Cl_3P{=}O$$

$$(PhO)_5P + R{-}NH_2 \xrightarrow[2\text{ h}]{150\,°C} (PhO)_3P{=}N{-}R + 2PhOH \quad (R = Ph;\ \text{m.p. } 254\,°C)$$

3. $$R_3P + N_3R' \xrightarrow[1\text{ h}]{\text{ether}} R_3P{=}N{-}R' + N_2$$

(R' = Ph, PhCO, Ph_3Si, Me_3Si, $R{-}SO_2$, etc.)

$R_3P{=}N{-}N{=}N{-}R'$ has been isolated as an intermediate in some of these reactions.

$$R_3P + N{\equiv}N^+{-}\bar{C}\langle^{R^1}_{R^2} \xrightarrow{Et_2O} R_3P{=}N{-}N{=}C\langle^{R^1}_{R^2}$$

(R = Ph, $R^1 = R^2$ = H; 81% m.p. 145 °C)

Table 30 – continued

4. $Ph_3P{=}CH{-}Ph + ArCH{=}N{-}Ph \xrightarrow{\text{benzene}} Ph_3P{=}NPh + ArCH{=}CHPh$

5. Substitution of halogens on phosphorus is also possible.

$$Cl_3P{=}N{-}COR + 3R'O^-Na^+ \longrightarrow (R'O)_3P{=}N{-}COR + 3NaCl$$

$(R' = R = Ph;\ m.p.\ 74\,°C)$

$$Cl_3P{=}N{-}SO_2Ar + 6NH_3 \longrightarrow (NH_2)_3P{=}N{-}SO_2Ar + 3NH_4^+Cl^-$$

4.3.2 *Reactions*

Until recently, the imidophosphoranes had been less extensively studied than phosphonium ylids. In their reactions they are closely related to the phosphonium ylids and exhibit two basic characteristics, nucleophilic nitrogen and ready elimination of phosphorus-containing compounds via small-membered ring transition states.

$$R_3P{=}N{-}R' \xrightarrow{HX} R_3P^+{-}NHR' \quad (4.148)$$

As with phosphonium ylids, treatment with acid leads to phosphonium salts (4.148). Imidophosphoranes are generally protonated less readily than are the analogous phosphonium ylids; conversely, the aminophosphonium salts (4.148) are more acidic than the equivalent phosphonium salts. For example, the salt (4.149) can be converted to the ylid (4.150) by treatment with triethylamine, while methyltriphenylphosphonium bromide (4.151) requires treatment with metal alkyls to form the corresponding ylid (4.152).

$$\underset{(4.149)}{Ph_3P^+{-}NH_2}\ Br^- \xrightarrow{Et_3N} \underset{(4.150)}{Ph_3P{=}N{-}H} + Et_3N^+H\ Br^-$$

$$\underset{(4.151)}{Ph_3P^+{-}CH_3}\ Br^- \xrightarrow{MeLi} \underset{(4.152)}{Ph_3P^+{-}C^-H_2} + LiBr + CH_4$$

$$Ph_3P{=}N{-}R \xrightarrow{H_2O} Ph_3P^+{-}NHR\ \ OH^- \longrightarrow Ph_3P{=}O + RNH_2$$

The imidophosphoranes are hydrolysed fairly readily, the conditions required being dependent on the substituents on the nitrogen atom. Presumably, the reaction proceeds via the corresponding phosphonium salt (cf. phosphonium ylids, p. 150), and the products are an amine and phosphine oxide.

Table 31 Reactions of imidophosphoranes

$$R_3P{=}N{-}R' + CH_3I \longrightarrow R_3P^+{-}N(CH_3){-}R' \quad I^-$$

$$R_3P{=}N{-}R' + PhCOCl \longrightarrow R_3P^+{-}N(CO{-}Ph)(R') \quad Cl^-$$

$$R_3P{=}N{-}R' + CO_2 \longrightarrow R_3P{=}O + R'N{=}C{=}O$$

$$R_3P{=}N{-}R' + SO_2 \longrightarrow R_3P{=}O + R'N{=}S{=}O$$

$$R_3P{=}N{-}R' + (CH_3)_2CO \longrightarrow R_3P{=}O + (CH_3)_2C{=}NR'$$

$$R_3P{=}N{-}R' + R''N{=}C{=}O \longrightarrow R_3P{=}O + R'N{=}C{=}NR''$$

$$R_3P{=}N{-}Ar + NOCl \longrightarrow R_3P{=}O + ArN^+{\equiv}N \quad Cl^-$$

$$R_3P{=}N{-}R' + MeOOC{-}C{\equiv}C{-}COOMe \longrightarrow R_3P^+{-}C^-(COOMe){-}C(COOMe){=}N{-}R'$$

$$R_3P{=}N{-}N{=}CR'_2 + PhCHO \longrightarrow R_3P{=}O + PhCH{=}N{-}N{=}CR'_2$$

$$R_3P{=}N{-}CO{-}R' \xrightarrow{\text{heat}} R_3P{=}O + R'C{\equiv}N$$

$$R_3P{=}N{-}N{=}CR'_2 + H_2O \longrightarrow \left[\begin{array}{c} R'_2C \\ HN \quad N \\ | \quad \| \\ HN \quad N \\ R'_2C \end{array}\right] \longrightarrow R'_2C{=}N{-}NH_2$$

The ylids can be alkylated (or acylated) at the nitrogen atom in the same way as phosphonium ylids, and the reactions with double and triple bonds are also

$$Ph_3P{=}NR + O{=}C(R'){-}R' \longrightarrow \left[\begin{array}{c} Ph_3P^+ \quad\; R \\ N \\ | \\ O^-{-}C{-}R' \\ R' \end{array}\right] \longrightarrow Ph_3P{=}O + R{-}N{=}CR'_2$$

(4.153)

analogous (see p. 157). Examples of reactions of this type can be found in Table 31. The analogue of the Wittig reaction (see p. 142) is thought to take place via a similar four-centre intermediate (4.153), but this has not been confirmed.

Imide intermediates are thought to be involved in the formation of carbodiimides (4.154) from isocyanates, using phosphine oxides as catalysts; this high-yielding reaction may take place via the mechanism shown in Scheme **4.9**. Support for this mechanism is available from experiments with phosphine oxides labelled with oxygen-18, which show a steady fall in oxygen-18 content as the reaction proceeds.

$$R_3P{=}O + R'N{=}C{=}O \rightleftharpoons \left[\begin{array}{c} R_3P{-}O \\ | \quad\ | \\ R'{-}N{-}C{=}O \end{array}\right]$$

$$\downarrow$$

$$R_3P{=}N{-}R' + CO_2$$

$$\downarrow R'N{=}C{=}O$$

$$R_3P^+{-}N{-}R' \quad O^-{-}C{=}N{-}R'$$

$$R_3P{=}O + R'N{=}C{=}NR' \longleftarrow$$

(4.154)

Scheme **4.9**

It is apparent, therefore, that imidophosphoranes and phosphonium ylids are very similar in their reactions, differing mainly in reactivity. Because imidophosphoranes are chemically more stable than phosphonium ylids, they can be prepared with a greater variety of substituents on the phosphorus atom. For example, a large range of *P*-halogen-substituted (4.155) imidophosphoranes are known, whereas the equivalent phosphonium ylids are normally too reactive for isolation. Alkoxy-substituted imides (4.156) are also known, and it is interesting that, in the salts (4.157) derived from these compounds, an Arbusov reaction (see p. 54) leads to products (4.158) containing P=O bonds rather than P=N bonds, which have a lower bond energy.

$$X_3P{=}N{-}R'$$

(4.155; X = Cl, Br, F)

$$(RO)_3P^+{-}NR^1R^2\ Br^- \longrightarrow (RO)_2P(=O){-}NR^1R^2 + RBr$$

(4.157) (4.158)

$$(RO)_3P{=}N{-}R'$$

(4.156)

Problems

4.1 Allyltriphenylphosphonium bromide (4.159) reacts with phenyllithium to produce an ylid (4.160) which gives a phosphonium salt (4.161) when treated with methyl chlorocarbonate. The phosphonium salt (4.161) is hydrolysed by aqueous sodium hydroxide to give crotonic acid, and it reacts with benzaldehyde to give the diene (4.162). Explain.

$$\underset{(4.159)}{Ph_3P^+\text{—}CH_2\text{—}CH\text{=}CH_2 \quad Br^-} \xrightarrow{LiPh} \underset{(4.160)}{[\text{ylid}]}$$

$$\underset{(4.160)}{[\text{ylid}]} \xrightarrow{ClCOOMe} \underset{(4.161)}{[\text{phosphonium salt}]}$$

$$CH_3\text{—}CH\text{=}CH\text{—}COOH \xleftarrow{OH^-} \underset{(4.161)}{[\text{phosphonium salt}]}$$

$$\underset{(4.161)}{[\text{phosphonium salt}]} \xrightarrow{PhCHO} \underset{(4.162)}{PhCH\text{=}CH\text{—}CH\text{=}CH\text{—}COOMe}$$

[Bestmann and Schulz (1964).]

4.2 Write reasonable mechanisms for the following reactions.

(a) $$Ph_3P^+\text{—}\bar{C}R^1R^2 + R\text{—}C\equiv N^+\text{—}O^- \longrightarrow [\text{adduct}] \xrightarrow{\text{heat}} R^1R^2C\text{=}C\text{=}NR + \text{(azirine: ring of N, C(R), C(R}^1\text{)(R}^2\text{), with C=N)}$$

[Bestmann and Kunstmann (1969); Huisgen and Wulff (1969).]

(b) $$(Me_2CHO)_3P\text{=}N\text{—}N\text{=}C(Me)_2 \longrightarrow (Me_2CHO)_2P(\text{=}O)\text{—}NH\text{—}N\text{=}C(Me)_2 + MeCH\text{=}CH_2$$

[Poshkus and Horweh (1962).]

(c) $$R_3P\text{=}N\text{—}Ph + MeOOC\text{—}C\equiv C\text{—}COOMe \longrightarrow R_3P\text{=}C(COOMe)\text{—}C(COOMe)\text{=}N\text{—}Ph$$

[Brown, Cookson, Stevens, Mak and Trotter (1964).]

(d) $$8R_3P^+\text{—}C^-HR' + S_8 \longrightarrow 8R_3P\text{=}S + 4R'CH\text{=}CHR'$$

[Magerlein and Meyer (1970). See also Schönberg, Brosowski and Singer (1962) and Staudinger and Meyer (1919).]

(e) $R_3P + OH^- \xrightarrow[N_2]{H_2O} R_3P{=}O + H_2 + OH^-$,

where R_3P is a water-soluble phosphine.
[Bloom, Buckler, Lambert and Merry (1970).]

(f) 1-bromo-2-fluorobenzene $\xrightarrow[THF]{Mg}$ [benzyne] $\xrightarrow{Ph_2PMe}$ [X] $\xrightarrow{\text{cyclohexanone}}$ $Ph_3P{=}O$ + methylenecyclohexane ($=CH_2$)

[Seyferth and Burlitch (1963).]

(g) Cyclic phosphonium salt (P+ bearing Me and Ph, ring carbons bearing Me groups), I⁻ $\xrightarrow[\text{(ii) MeI}]{\text{(i) BuLi}}$ Me₂C(–CH(Me)–CMe₂Ph)–P⁺(Bu)(Me)–Me I⁻

[Corfield, Harger, Shutt and Trippett (1970).]

(h) $Ph_3P^+{-}CH_2CH_2{-}P^+Ph_3$ $2Br^-$ $\xrightarrow{PhLi}$ Ph_3P + unidentified products

$Ph_3P^+{-}CH_2CH_2{-}P^+Ph_3$ $2Br^-$ $\xrightarrow{\text{piperidide } N^-Li^+}$ $Ph_3P + Ph_3P^+{-}CH_2CH_2{-}N$(piperidine) Br^-

[Wittig, Eggers and Duffner (1958).]

(i) $Ph_3P^+{-}C^-H(CH_2)_3COOEt \longrightarrow$ 2-oxocyclopentylidene ylide (ring C⁻ bearing P^+Ph_3)

[Bergelson, Vaver, Barsukov and Schemyakin (1963); House and Babad (1963).]

(j) $Ph_3P^+{-}CH{=}CH_2$ Br^- + salicylaldehyde sodium salt (CHO, O^-Na^+) $\longrightarrow$ 2H-chromene + $Ph_3P{=}O$

[Schweizer (1964); Schweizer and Light (1964).]

4.3 Write mechanisms for the reactions in Table 28 (p. 157).

4.4 Explain the routes taken in the following hydrolyses.

(a) [dibenzophospholium salt, P^+ bearing R and CH_2I] $\xrightarrow{OH^-}$ [ring-expanded phosphine oxide: P(=O)R bridging biphenyl via CH_2]

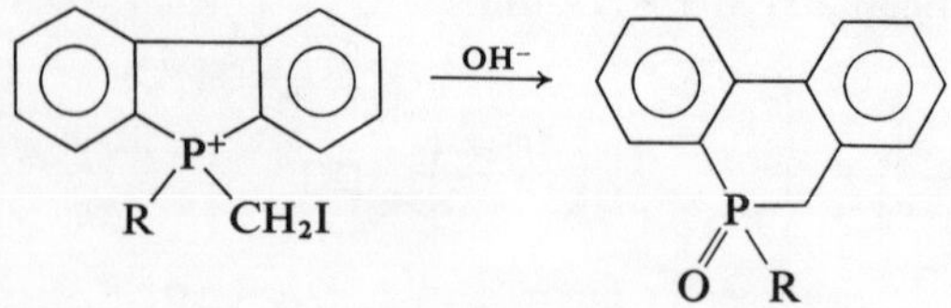

(b) [four-membered ring P^+—CH_2I, R] $\xrightarrow{OH^-}$ [five-membered ring P(=O)R]

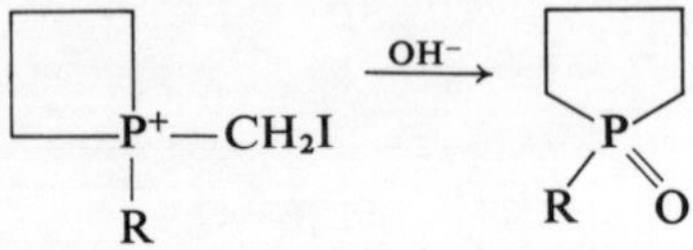

(c) [five-membered ring P^+, R, CH_2I] $\xrightarrow{OH^-}$ [six-membered ring P(=O)R]

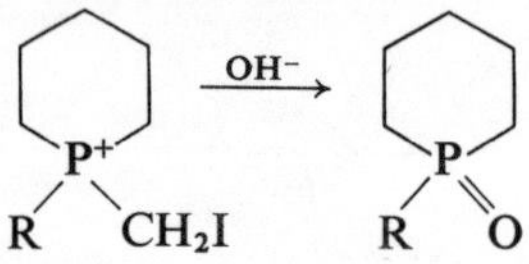

(See Chapter 4, pp. 131–3.)

4.5 Refluxing 2-hydroxyethyltriphenylphosphonium salts (4.163) with aqueous sodium hydroxide leads to the formation of β-hydroxyethyldiphenylphosphine oxide (4.164) and benzene. However, under similar conditions, the ester salt (4.165) undergoes a Hofmann elimination to produce vinyl acetate and triphenylphosphine. Explain.

$$\underset{(4.163)}{Ph_3P^+CH_2CH_2OH} \xrightarrow[H_2O]{OH^-} \underset{(4.164)}{Ph_2\overset{\overset{O}{\|}}{P}CH_2CH_2OH}$$

$$\underset{(4.165)}{Ph_3P^+CH_2CH_2OCOMe} \xrightarrow[H_2O]{OH^-} Ph_3P + CH_2{=}CH{-}OCOMe$$

4.6 Sodium diethylphosphate (4.166) reacts with *cis*-stilbene oxide (4.167) to give *trans*-stilbene, while a similar reaction with *trans*-stilbene oxide (4.168) gives *cis*-stilbene. Explain.

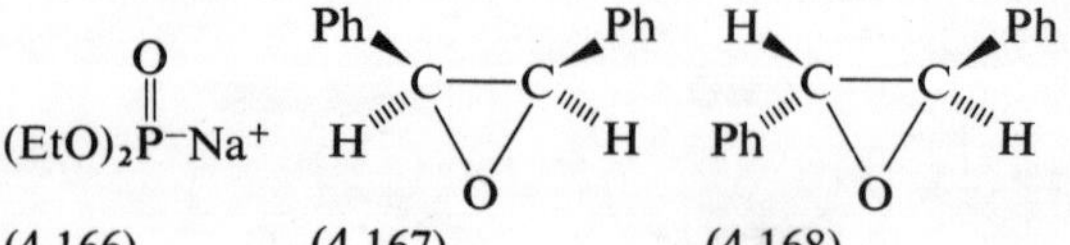

$$\underset{(4.166)}{(EtO)_2\overset{\overset{O}{\|}}{P}{}^-Na^+}$$

(4.167) cis-stilbene oxide: Ph, H on one C; Ph, H on the other C (epoxide O)

(4.168) trans-stilbene oxide: H, Ph on one C; Ph, H on the other C (epoxide O)

4.7 The n.m.r. spectrum of the aldehyde-stabilized phosphonium ylid (4.169) in $CDCl_3$ solution at $-12\,°C$ shows, amongst other absorptions, a doublet of doublets at $1{\cdot}72\ \tau$ ($J_1 = 4$ Hz, $J_2 = 11$ Hz) and a doublet of doublets at $1{\cdot}08\ \tau$ ($J_3 = 39$ Hz, $J_4 = 3{\cdot}5$ Hz). The ratio of these two groups of peaks, obtained from integration, is 47:53. As the temperature of this solution is increased, these resonances broaden and, at 80 °C, they coalesce to a broad unsymmetrical singlet. Lowering the temperature of the solution reverses the changes. Explain these observations.

$Ph_3P^+—C^-H—C(=O)H$
(4.169)

[Snyder and Bestmann (1970). However, see also Wilson and Tebby (1970) and Filleux-Blanchard and Martin (1970).]

4.8 The ylid (4.170) reacts with hexafluoroacetone at 40 °C in diglyme to give a stable crystalline compound (4.171) showing the following characteristics: molecular weight (by osmometry) 700 and 706; ^{31}P n.m.r. in CH_2Cl_2 showed doublets of equal intensity at $-7{\cdot}3$ p.p.m. and $+54$ p.p.m. ($J_{pp'} = 47$ Hz); ^1H n.m.r. showed only a broad absorption between $+2$ and $+3{\cdot}5\ \tau$.

(4.171) was not alkylated by methyl iodide at 40 °C after 24 hours, but did react with acid to give compound (4.172) which showed only a moderately broad singlet at -22 p.p.m. in its ^{31}P n.m.r. spectrum. When (4.171) was heated to above 110 °C, and the reaction product was treated with hydrogen chloride gas, triphenylphosphine oxide and a phosphonium salt (4.173) were obtained. (4.173) showed a singlet at $-17{\cdot}3$ p.p.m. in its ^{31}P n.m.r. and a doublet ($J_{PH} = 8{\cdot}5$ Hz) at $+0{\cdot}9\ \tau$ together with aryl multiplets in its ^1H n.m.r.

Suggest structures for (4.171, 4.172 and 4.173) which are compatible with the information given.

$$Ph_3P^+—C^{2-}—P^+Ph_3 + (CF_3)_2CO \xrightarrow{\text{diglyme}} (4.171) \xrightarrow{\text{HCl}} (4.172)$$
(4.170)

(4.171) → (i) >110 °C (ii) HCl → Ph_3PO + (4.173)

[Birum and Matthews (1967).]

Chapter 5
Phosphorus in Biological Processes

5.1 Energy transfer in living systems

All living organisms consume energy and possess processes for its transfer, conversion, etc. However, different organisms use energy in different ways. Mammals, for example, use a considerable amount of energy in muscular action and temperature control. On the other hand, bacteria tend to be immobile, and use less energy in environmental control than in synthesis, since their very existence depends on extremely rapid reproduction (some bacteria can synthesize their own weight of 'body' material in fifteen minutes). Although the final uses of energy differ in these ways, the basic processes by which it is stored and transported are common to all living systems.

The ultimate energy source in all organisms is the intracellular breakdown of food molecules. This energy is conserved in adenosine triphosphate (ATP)

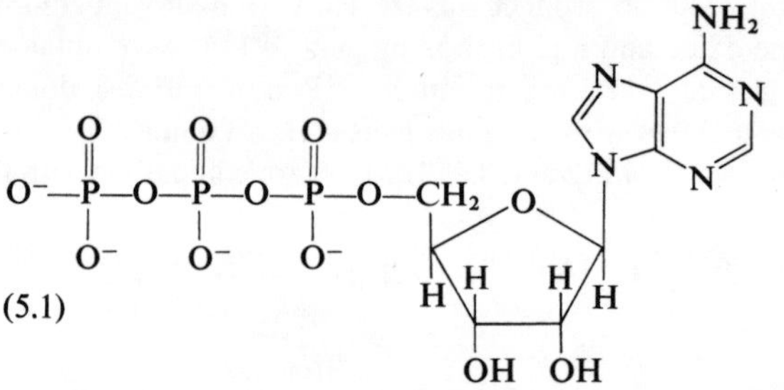

(5.1)

$$\text{ADP} + PO_4^{3-} + \text{energy} \rightarrow \text{ATP}$$

Scheme **5.1** $(29{\cdot}3\ \text{kJ}\ (\text{mol ADP})^{-1})$

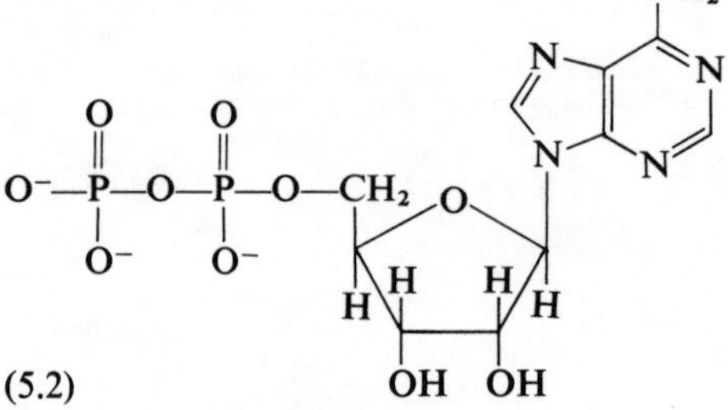

(5.2)

(5.1) by the synthesis of this compound from adenosine diphosphate (ADP) (5.2) and phosphate (PO_4^{3-}) according to Scheme **5.1**. This 'stored' energy can then be transferred as ATP to sites where energy is needed to carry out various types of work (e.g. biosynthetic, mechanical). By incorporating the conversion of ATP to ADP into the energy-requiring process, the energy stored in ATP can be utilized.

Phosphate source + ADP —(Energy from food; e.g. respiration)→ ATP

ATP ↓ (→ work) ADP

ADP → (returns to) Phosphate source + ADP

How this conversion process makes energy available for work is best understood in terms of free energy. Consider the hydrolysis of ATP to ADP *in vitro* in the absence of other components. This reaction (Scheme **5.2**) shows an overall free-energy change (ΔG) of $-29{\cdot}3\ \mathrm{kJ\,mol^{-1}}$ ($-7\ \mathrm{kcal\,mol^{-1}}$). (The negative charges signify the ionized state of the molecules at the pH of most biological systems.) The negative sign indicates an *evolution* of energy. Thus, if the ATP to ADP conversion can be incorporated into some other chemical process, then that process is able to use an additional 29·3 kJ of energy for every mole of ATP converted. This, in simple terms, is how ATP is used as an energy source.

$$ATP^{4-} + H_2O \longrightarrow ADP^{3-} + O^-\!\!-\!\overset{\overset{\displaystyle O}{\|}}{\underset{\underset{\displaystyle O^-}{|}}{P}}\!-\!O^- + 2H^+$$

Scheme **5.2**

Since the ATP to ADP conversion gives out energy, then the conversion of ADP and phosphate into ATP (the reverse process) will require the input of an equal amount of energy (see Scheme **5.3**). (The energy required comes from various processes, for example respiration, pp. 174–6.)

$$ADP^{3-} + O^-\!\!-\!\overset{\overset{\displaystyle O}{\|}}{\underset{\underset{\displaystyle O^-}{|}}{P}}\!-\!O\!-\!R \longrightarrow ATP^{4-} + R\!-\!O^-$$

$$\Delta G = +\,29{\cdot}3\ \mathrm{kJ\,mol^{-1}}\ (+\,7\ \mathrm{kcal\,mol^{-1}})$$

Scheme **5.3**

In the cell, ATP is continually being hydrolysed to obtain energy, and then resynthesized. ATP acts as an *energy carrier*, and its versatility in reaction enables it to transfer energy to a wide range of processes. Though there are many other similar phosphates involved in metabolic processes, the main flow of energy within the cell is controlled by the ATP → ADP interconversion cycle.

5.1.1 *Regeneration of ATP from ADP*

In reactions in which the free energy of the products is lower than the free energy of the reagents, the extra energy is usually given out as heat. However, in a cell, where these reagents have usually been synthesized by processes involving considerable expenditure of energy, evolution of heat would be largely wasted energy. To overcome this problem, the cell involves such reactions in the conversion of ADP to ATP. Since this conversion requires an input of energy, the latter is conserved for use in other processes. In animal cells there are two main ways in which this conversion takes place.

(a) *Anaerobic ADP to ATP conversion.* Cells can be divided into two types; those which can live in the presence of oxygen (aerobic cells) and those which can only live in the absence of oxygen (anaerobic cells). There are relatively few examples of anaerobic cells but, since many cells can modify their metabolic processes so as to survive either condition, anaerobic processes are quite important.

All cells which can survive in the absence of oxygen must be able to convert ADP to ATP by a process which does not involve oxygen; otherwise, once they had converted all their ATP to ADP, they would stop functioning. This would happen very quickly: a cell deprived of a method of converting ADP to ATP will use up all its available ATP in a matter of seconds.

The energy required to convert ADP to ATP anaerobically is obtained from the conversion of glucose to lactic acid. This process is very complex but basically the conversion involves ADP and phosphate to form ATP. An overall equation can be written (Scheme **5.4**),

$$\text{Glucose} + 2\,O^{-}\!-\!\overset{\overset{\displaystyle O}{\|}}{\underset{\underset{\displaystyle O^{-}}{|}}{P}}\!-\!O^{-} + 2\,\text{ADP} \longrightarrow 2\,\text{lactate} + 2\,\text{ATP}$$

Scheme **5.4**

but the actual process involves eleven steps, each one involving a specific enzyme.

(b) *Aerobic ADP to ATP conversions.* The ADP to ATP conversion process utilizing oxygen is known as 'oxidative phosphorylation' and involves two very

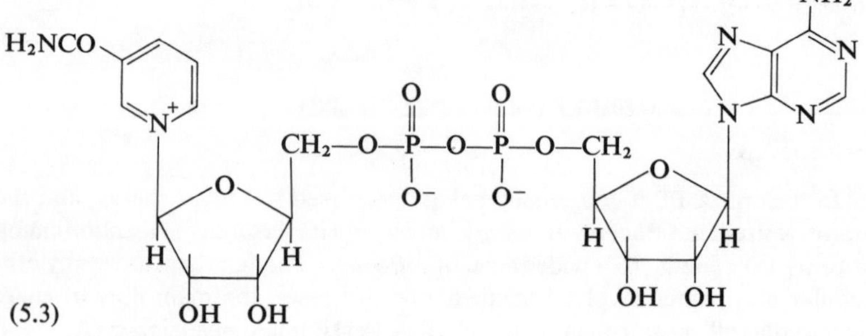

(5.3)

important phosphate esters, nicotinamide adenine dinucleotide (NAD) (5.3), and coenzyme A (5.4).

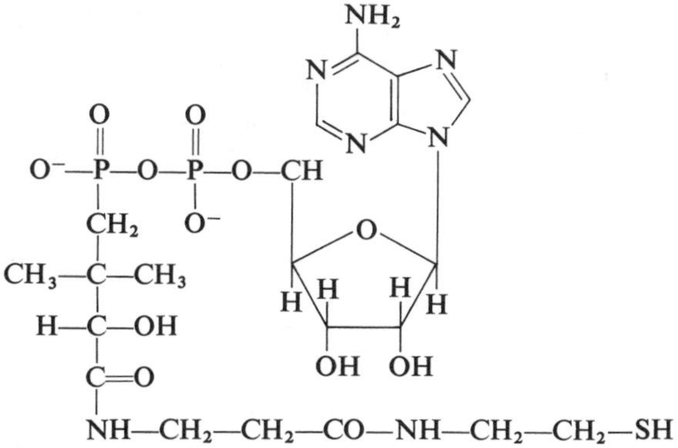

$$
\begin{array}{l}
O^{-}\text{—}P(=O)\text{—}O\text{—}P(=O)(O^{-})\text{—}O\text{—}CH \ (\text{ribose–adenine; } H, H, H, H; \ OH \ OH) \\
\quad\ \ CH_2 \\
CH_3\text{—}C\text{—}CH_3 \\
\quad H\text{—}C\text{—}OH \\
\qquad C{=}O \\
\qquad NH\text{—}CH_2\text{—}CH_2\text{—}CO\text{—}NH\text{—}CH_2\text{—}CH_2\text{—}SH
\end{array}
$$

(5.4)

NAD is one of the cell's electron carriers and is therefore intimately concerned in any oxidative process. Cells convert acetic acid and oxygen into carbon dioxide and water in a complex series of equilibria known as Krebs's tricarboxylic acid cycle; the acetic acid is introduced into the process as the acetyl group in *S*-acetyl coenzyme A (coenzyme A—S—CO—CH_3). During the process, NAD (5.3) undergoes a two-electron reduction at the pyridine ring to give the dihydropyridine form (5.5).

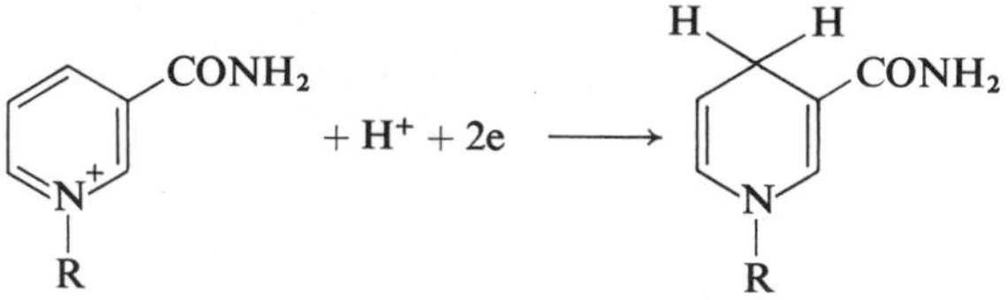

$$\text{(pyridinium, } N^{+}\text{—R, } CONH_2) + H^{+} + 2e \longrightarrow \text{(1,4-dihydropyridine, } H, H, \ N\text{—R, } CONH_2)$$

(5.3) (5.5)

The reduced form of NAD is then reoxidized by an oxygen atom in a process which incorporates ADP and phosphate, the overall equation being

$$\underset{\text{(reduced form)}}{\text{NAD}} + \tfrac{1}{2}O_2 + 3\,\text{ADP} + 3PO_4^{3-} \longrightarrow \text{NAD} + 3\text{ATP} + 4H_2O,$$

although the actual process is much more complex than this equation suggests. The oxidation of the dihydropyridine form of NAD (5.5) is accompanied by a large decrease in free energy ($\Delta G = -218$ kJ mol^{-1} (-52 kcal mol^{-1})) and each mole of ADP involved in this reaction is able to utilize 29·3 kJ (7 kcal) of this energy in conversion to ATP. Since three molecules of ADP are converted to ATP during the oxidation of each molecule of reduced NAD (5.5), then

87·9 kJ (21 kcal) of the total energy available are used in the conversion and hence saved for use in other processes requiring an input of energy.

The conversion of ADP to ATP by respiration (the aerobic process) is much more efficient in terms of 'fuel' consumption than the anaerobic conversion process discussed earlier. The efficiency of the aerobic energy-conversion process has been calculated to be as high as 60 per cent. It is interesting to note that the actual rate of respiration in animals is controlled by the requirements of the ADP to ATP conversion in maintaining an acceptable concentration of ATP in the cell. For example, as muscular activity uses up ATP (see below), the rate of respiration increases to the extent necessary to replace the loss. However, in certain circumstances, anaerobic activity in the muscle cells also increases, leading to a build up of lactic acid which causes the discomfort often associated with extreme muscular activity.

5.1.2 *Use of the energy stored in ATP in metabolic processes*

(a) *Muscular action.* In order to derive energy from ATP for use in a process such as muscular action, the ATP must be hydrolysed to ADP and phosphate, a reaction with $\Delta G = -29{\cdot}3\ \mathrm{kJ\ mol^{-1}}$ ($-7\ \mathrm{kcal\ mol^{-1}}$). However, to use this energy, the mechanism of muscular action must include the conversion of ATP to ADP in its pathway.

About 50 per cent of muscle fibre consists of a protein called myosin. Isolated samples of myosin show enzymic activity in the ATP to ADP conversion, that is, they catalyse this reaction. Other similar constituents of muscle show this same enzymic behaviour towards the conversion and, during the hydrolysis, the *shapes* of these proteins probably alter. These changes in shape (probably in a conformational rather than a structural sense) constitute the contraction and relaxation of muscle fibre that is required for muscular action. These changes will obviously require energy, and this is provided by the ATP hydrolysis process.

Since muscles are frequently expected to carry out rapid and repetitive action, it is important that there should be some easy way of converting ADP back to ATP to provide energy for further activity. We have already seen that respiration can do this, but, in phases of extreme activity, additional energy may be supplied by phosphocreatine (5.6). High concentrations of phosphocreatine are present in both muscle and nerve cells, where it can act as a reserve store of energy by phosphorylating ADP (formed by muscular action) back to ATP. The other product of this phosphorylation is creatine (5.7), which can be phosphorylated back to phosphocreatine by respiration during periods of muscle rest.

$$\mathrm{^{-}O{-}\overset{\overset{\displaystyle O}{\|}}{\underset{\underset{\displaystyle O^-}{|}}{P}}{-}\underset{\underset{\displaystyle H}{|}}{N}{-}\underset{\underset{\displaystyle NH}{\|}}{C}{-}\overset{\overset{\displaystyle CH_3}{|}}{N}{-}CH_2{-}COO^{-}} \qquad (5.6)$$

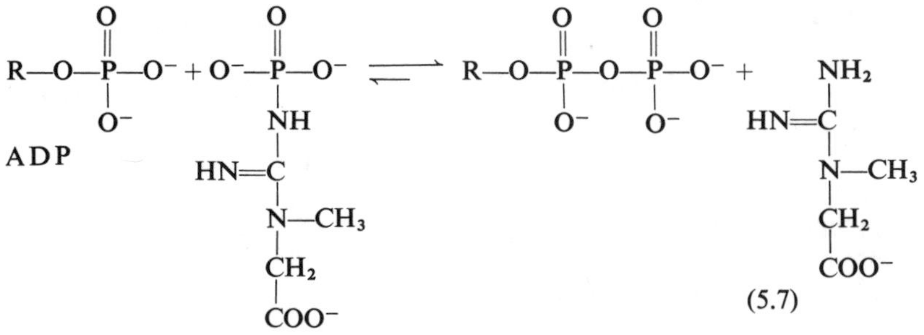

$$R{-}O{-}P(=O)(O^-){-}O^- + O^-{-}P(=O)(O^-){-}NH{-}C(=NH){-}N(CH_3){-}CH_2{-}COO^- \rightleftharpoons R{-}O{-}P(=O)(O^-){-}O{-}P(=O)(O^-){-}O^- + NH_2{-}C(=NH){-}N(CH_3){-}CH_2{-}COO^- \qquad (5.7)$$

ADP

Phosphate esters such as phosphocreatine are often known as *high-energy* phosphate esters because they have very large negative free energies of hydrolysis, $\Delta G = -44$ kJ mol^{-1} ($-10{\cdot}5$ kcal mol^{-1}), and the net result of an equilibrium with ADP will largely be the formation of ATP (i.e. the equilibrium will always tend to be to the right).

(b) *Biosynthesis.* In every cell there is a large number of complicated compounds, each of which must be synthesized by the cell from relatively simple starting materials. We shall study some more specific aspects of this when dealing with nucleic acids but, in general, these syntheses will require energy and the main source of this energy is ATP.

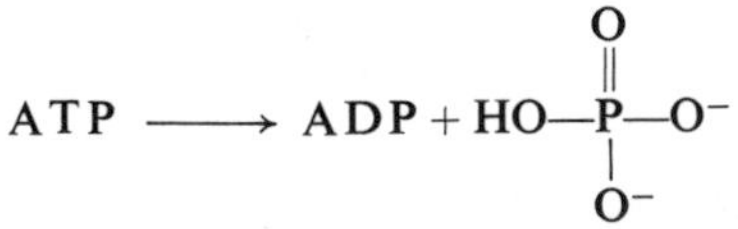

$$\text{ATP} \longrightarrow \text{ADP} + HO{-}P(=O)(O^-){-}O^-$$

$$\Delta G = 29{\cdot}3\,\text{kJ mol}^{-1}\ (-7\,\text{kcal mol}^{-1})$$

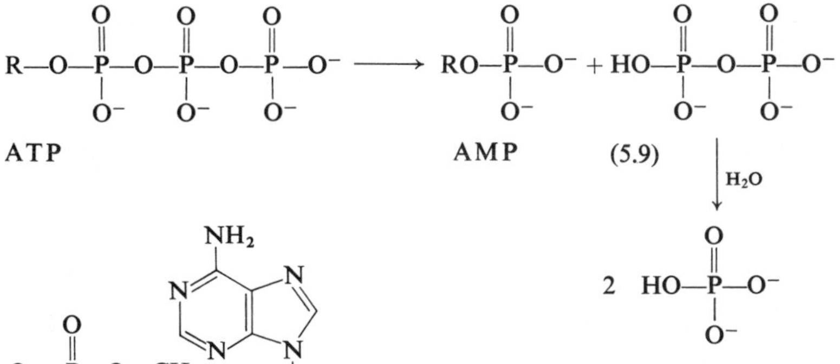

$$R{-}O{-}P(=O)(O^-){-}O{-}P(=O)(O^-){-}O{-}P(=O)(O^-){-}O^- \longrightarrow RO{-}P(=O)(O^-){-}O^- + HO{-}P(=O)(O^-){-}O{-}P(=O)(O^-){-}O^- \qquad (5.9)$$

ATP　　　AMP

$$\xrightarrow{H_2O}\ 2\ HO{-}P(=O)(O^-){-}O^-$$

$$O^-{-}P(=O)(O^-){-}O{-}CH_2{-}\text{(ribose, H, H, H, H, OH, OH)}{-}\text{adenine }(NH_2, N, N, N, N)$$

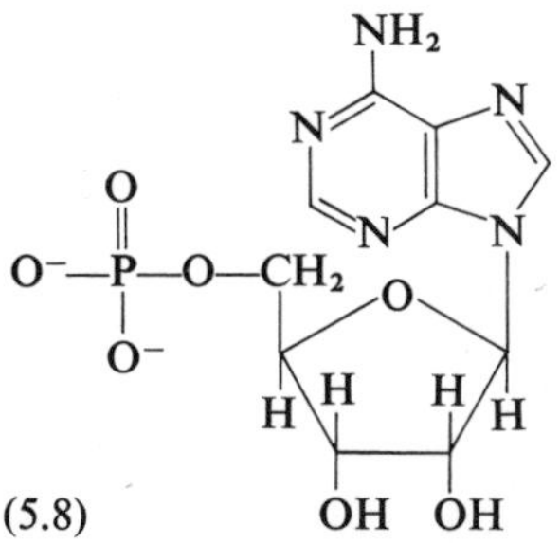

(5.8)

In earlier discussions, we have assumed that the ATP energy-releasing process involves decomposition to ADP and phosphate, a process which has a free-energy change of 29·3 kJ mol^{-1} (7 kcal mol^{-1}). However, there is an alternative mode of decomposition of ATP which makes available approximately twice as much energy. ATP can undergo hydrolysis at the first phosphate group, rather than at the second, to give adenosine monophosphate (AMP) (5.8) and pyrophosphate (5.9). The pyrophosphate can then undergo further hydrolysis, producing two molecules of phosphate. The overall free energy change is −58·6 kJ mol^{-1} (−14 kcal mol^{-1}), so twice as much energy is available from each molecule of ATP in this process as in the conversion to ADP. This larger free-energy change is especially important in biosynthesis, where the reaction product is required and so the *yield* needs to be as high as possible.

Let us consider glycogen synthesis as an example of how the cell utilizes the energy of ATP in biosynthesis. Glycogen (5.10) is a polysaccharide consisting of a long chain of glucose units. The number of units in each molecule varies, but the molecular weight of glycogen can be as high as 8000000.

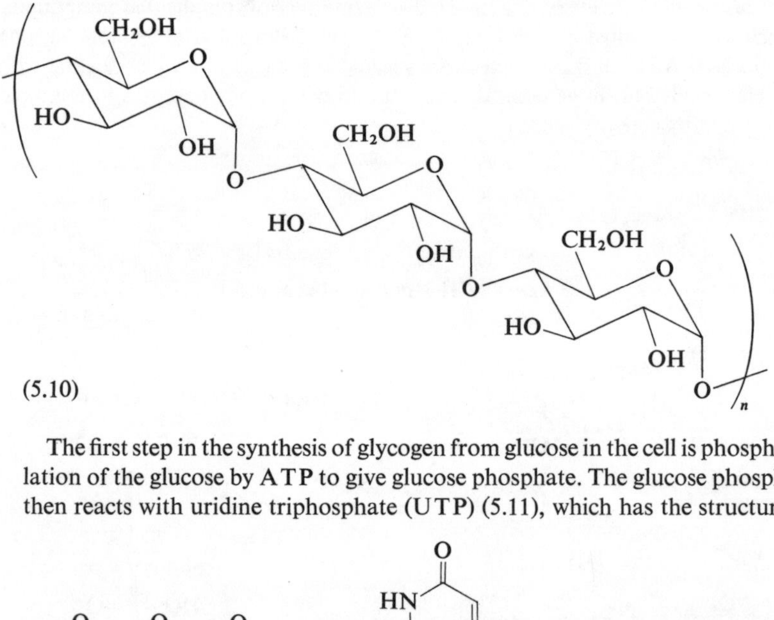

(5.10)

The first step in the synthesis of glycogen from glucose in the cell is phosphorylation of the glucose by ATP to give glucose phosphate. The glucose phosphate then reacts with uridine triphosphate (UTP) (5.11), which has the structure of

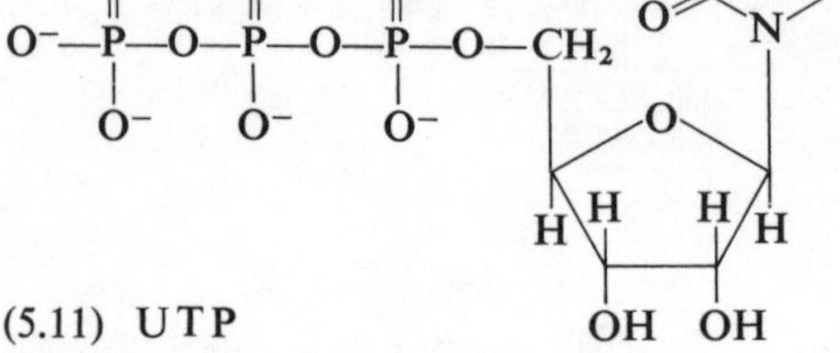

(5.11) UTP

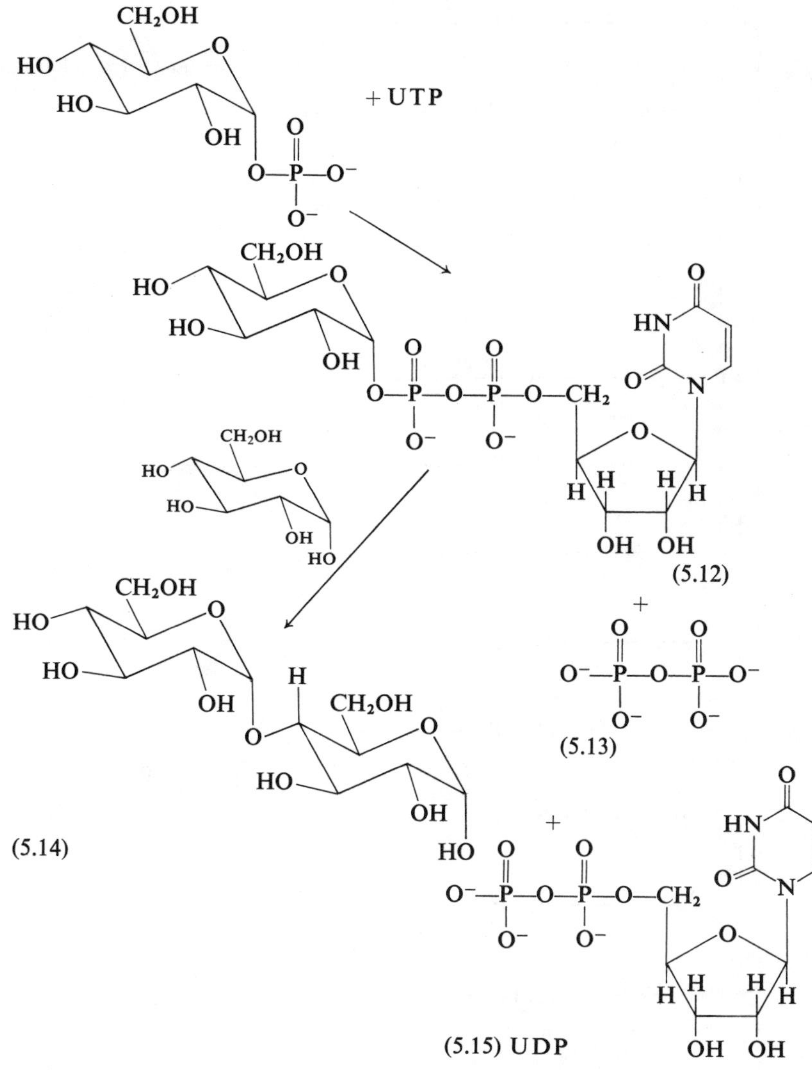

+UTP

(5.12)

(5.13)

(5.14)

(5.15) UDP

ATP but with the adenine base replaced by uracil, to give uridine diphosphoglucose (5.12) and pyrophosphate (5.13). The final step is the reaction between uridine diphosphoglucose (5.12) and glucose to give a disaccharide (5.14) and uridine diphosphate (UDP) (5.15). This process is continually repeated, one glucose unit being added to the chain each time. The uridine diphosphate (UDP) is converted by ATP back to the triphosphate (UTP) for use in further steps,

$$\mathrm{UDP + ATP \longrightarrow UTP + ADP.}$$

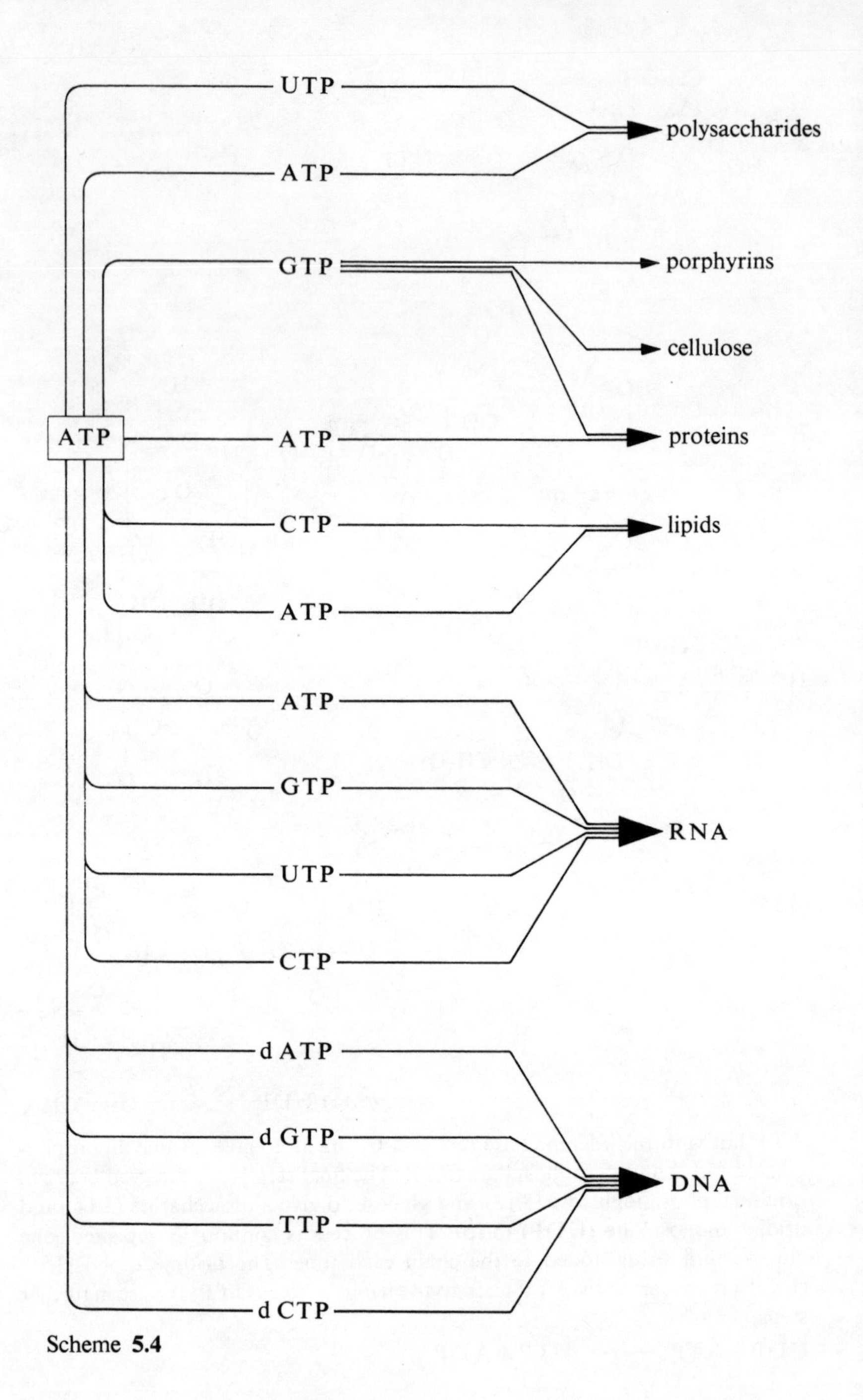
UTP
ATP
polysaccharides
GTP
porphyrins
cellulose
ATP
ATP
proteins
CTP
lipids
ATP
ATP
GTP
RNA
UTP
CTP
d ATP
d GTP
DNA
TTP
d CTP

Scheme **5.4**

In fact there are a large number of closely related di- and tri-phosphates present in the cell; the essential purpose of these is to energize various steps in the biosynthesis of cell components. The structures of these triphosphates are given in Table 32 and an outline of their use is given in Scheme **5.4**. They can be seen to fall into two groups, those in which the carbohydrate molecule is a ribose unit, and those in which it is a deoxyribose unit. These compounds are known as nucleoside triphosphates, and it is from units similar to these that the nucleic acids are built up: ribonucleic acids (RNA) from the ribose-based nucleosides, and deoxyribonucleic acids (DNA) from the deoxyribose-based units (see pp. 187–96).

ATP is so important because it is the central energy source; it derives its own energy from basic processes such as respiration and, by phosphorylation of diphosphates, it can transfer this energy to every triphosphate shown in Scheme **5.4**.

Table 32

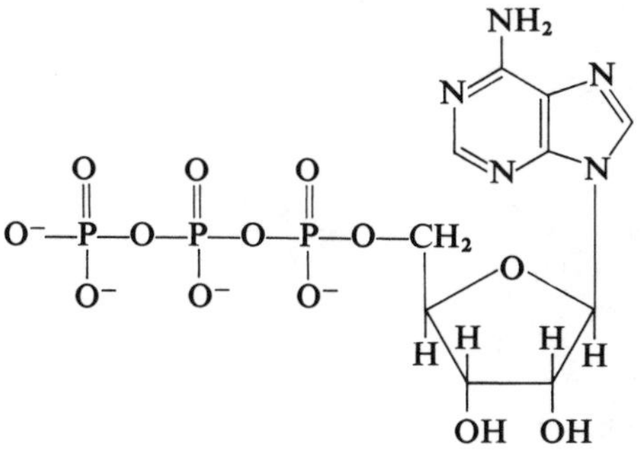

Compound	Structure
adenosine triphosphate (ATP)	O^-—P(=O)(O^-)—O—P(=O)(O^-)—O—P(=O)(O^-)—O—CH_2—[ribose ring: O; H, H, H, H; OH, OH]—adenine (NH_2; N, N, N, N)
deoxyadenosine triphosphate (dATP)	O^-—P(=O)(O^-)—O—P(=O)(O^-)—O—P(=O)(O^-)—O—CH_2—[deoxyribose ring: O; H, H, H, H; OH, H]—adenine (NH_2; N, N, N, N)

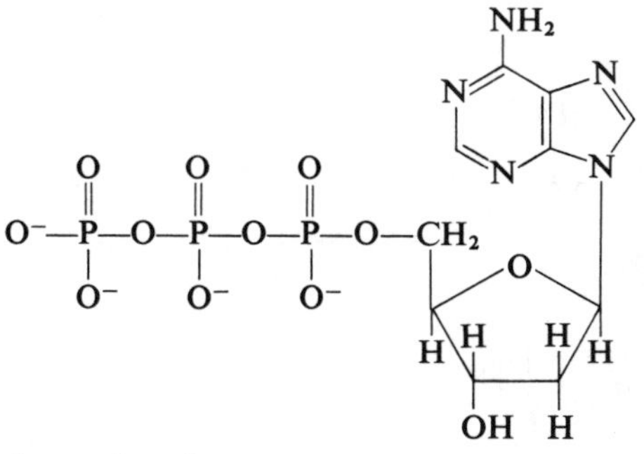

Table 32 – *continued*

guanosine
triphosphate
(GTP)

deoxyguanosine
triphosphate
(dGTP)

uridine
triphosphate
(UTP)

Table 32 – *continued*

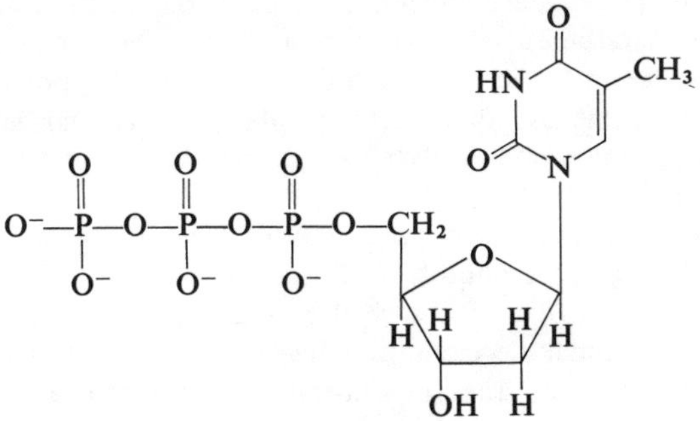

thymidine
triphosphate
(TTP)

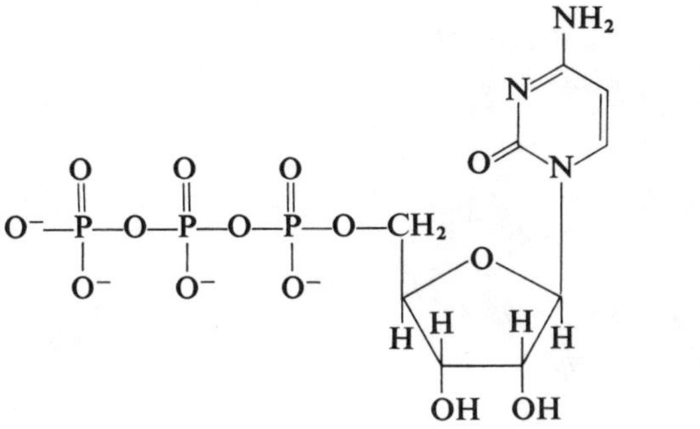

cytidine
triphosphate
(CTP)

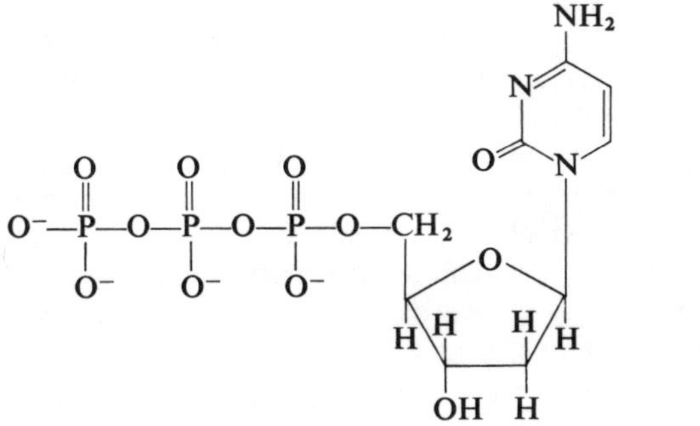

deoxycytidine
triphosphate
(dCTP)

5.1.3 *Photosynthesis*

ATP acts as the central energy-transfer agent in every cell, but the ways in which this energy is used differ with the type of cell. For example, plants have no muscles and so plant cells do not use ATP to produce muscular action. Methods of generation of ATP from ADP also differ between plant and animal cells. While animal cells use respiration and/or glycolysis (p. 174) for this purpose, plant cells use photosynthesis.

The overall chemical process which constitutes photosynthesis is the conversion of carbon dioxide into glucose photolytically (i.e. by the use of light). In the first step of this process, light is absorbed by chlorophyll (5.16) and promotes some of the chlorophyll electrons to a higher energy level. These excited electrons are then transferred from the chlorophyll to electron-carrying enzymes which use the extra energy in a variety of ways (including the synthesis of ATP from ADP and phosphate) before returning the electrons to the chlorophyll molecule. The electrons can then be re-excited by light and the process repeated. The synthesis of ATP in this way is known as photophosphorylation.

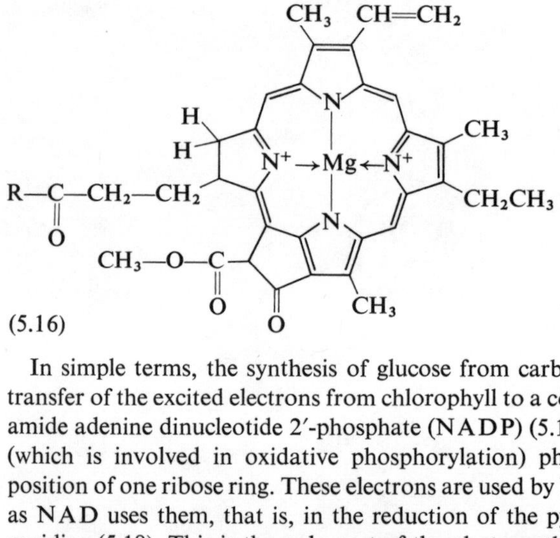

(5.16)

In simple terms, the synthesis of glucose from carbon dioxide involves the transfer of the excited electrons from chlorophyll to a compound called nicotinamide adenine dinucleotide 2′-phosphate (NADP) (5.17); this is simply NAD (which is involved in oxidative phosphorylation) phosphorylated in the 2′ position of one ribose ring. These electrons are used by NADP in the same way as NAD uses them, that is, in the reduction of the pyridine ring to dihydropyridine (5.18). This is the only part of the photosynthetic pathway of glucose synthesis that needs light, and the plant can continue to synthesize glucose during the night from starting-materials accumulated during the day.

The carbon dioxide absorbed by the plant reacts with the diphosphate (5.19) which is derived from ribose, to give two molecules of the phosphoglyceric ester (5.20), which is converted into glucose by ATP and the dihydropyridine form of NADP.

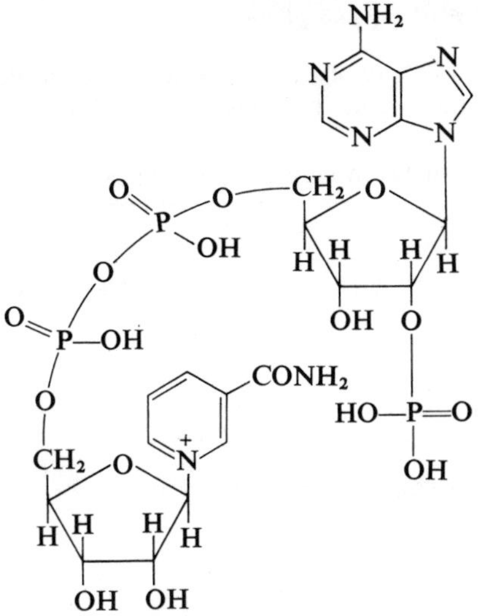

NADP

(5.17)

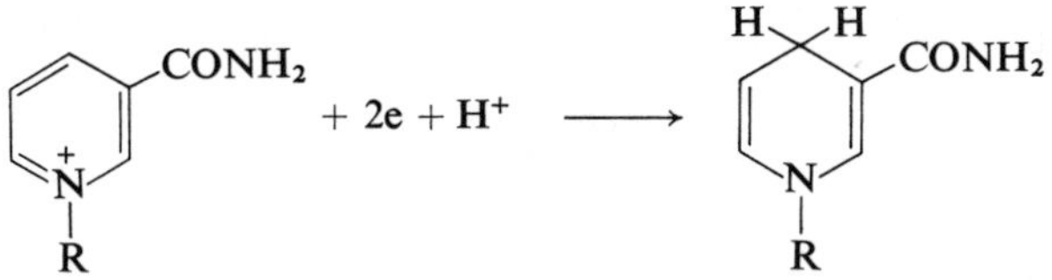

NADP $+ 2e + H^+ \longrightarrow$ (5.18)

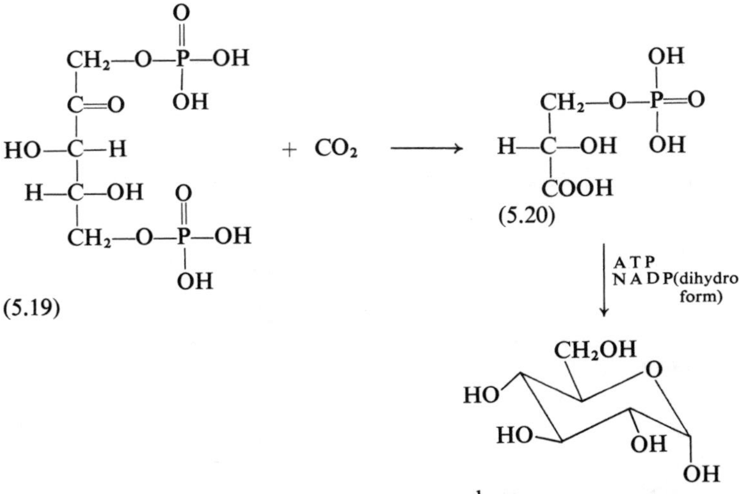

(5.19) $+ CO_2 \longrightarrow$ (5.20)

ATP
NADP(dihydro form)

glucose

5.1.4 *Other biosynthetic routes involving phosphates*

We have so far looked at what we might call 'general' or 'primary' processes. For example, photosynthesis occurs in all plants, respiration takes place in all

Table 33 Examples of organic compounds involved in secondary processes

Fatty acids

$CH_3(CH_2)_7CH{=}CH(CH_2)_{11}COOH$
eracic acid (*cis*)
brassidic acid (*trans*)

$CH_3(CH_2)_7CH{=}CH(CH_2)_7COOH$
oleic acid

Alkaloids

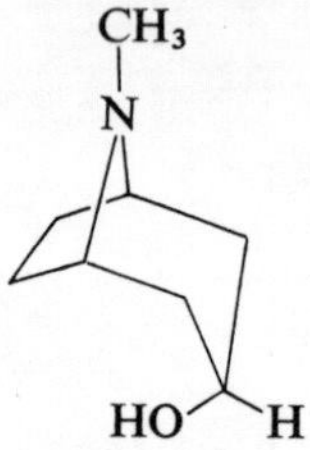

tropine

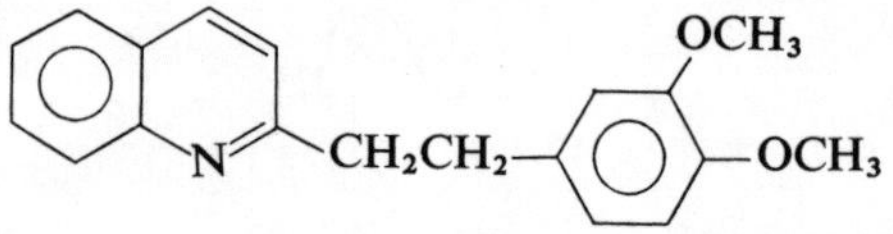

galipine

Terpenes

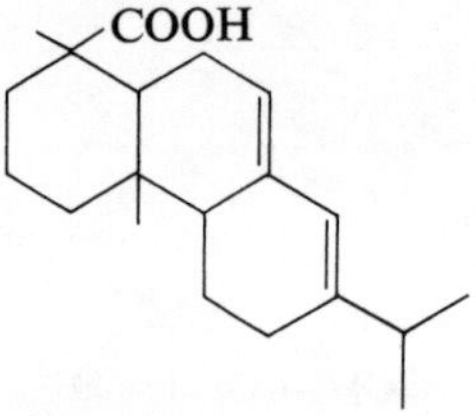

abietic acid

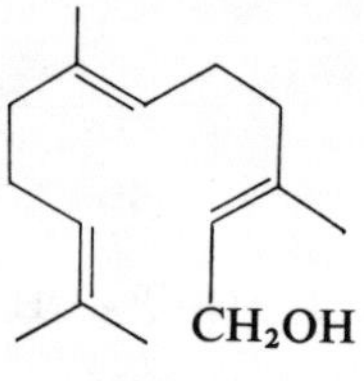

farnesol

Carotenoids

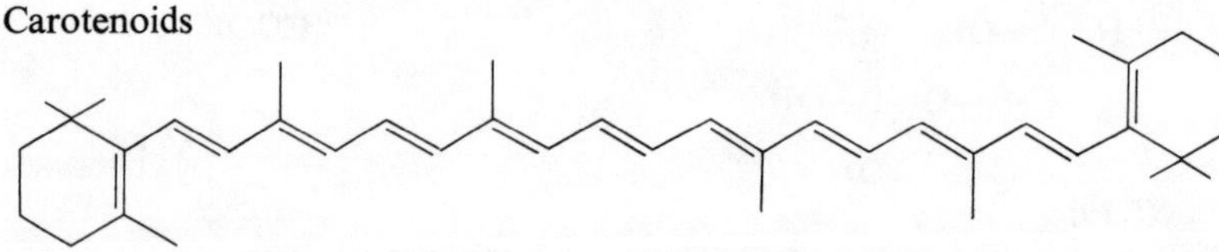

β-carotene

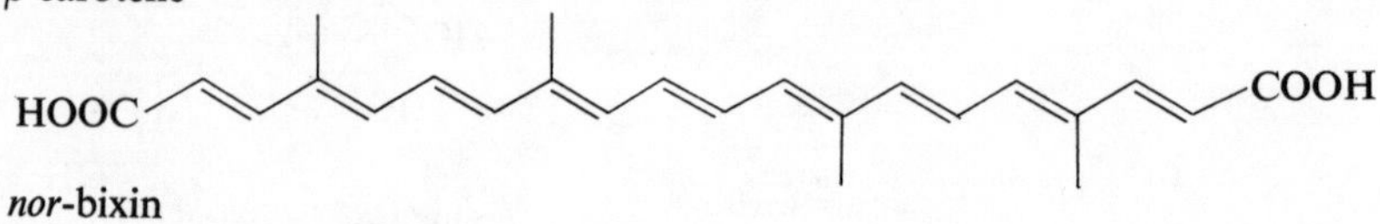

nor-bixin

Table 33 – continued

Flavonoids

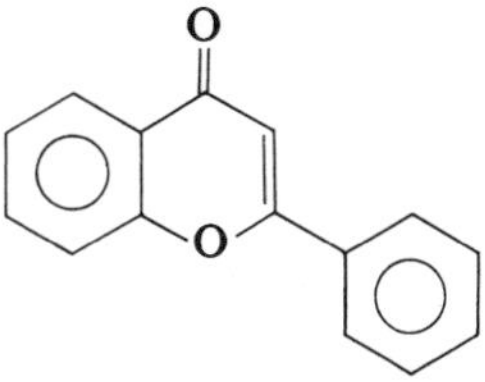

O OH HO O OH OH

flavone 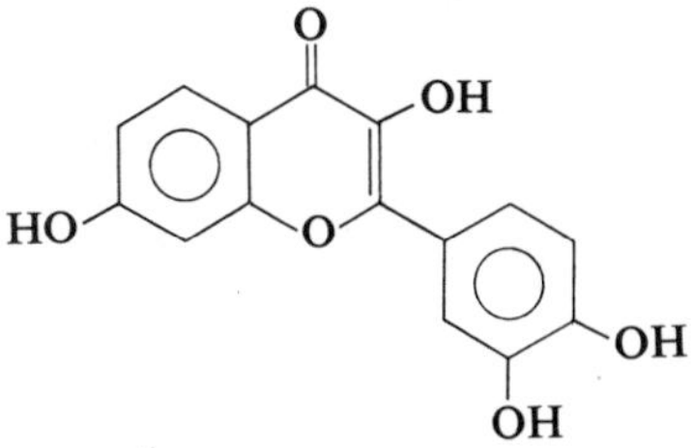 quercetin

iso-Flavonoid

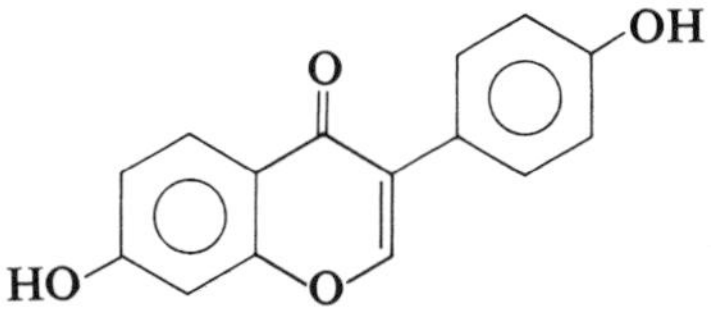

O OH HO O

diadzein

aerobic animal cells, and **ATP** to **ADP** conversions are the primary processes by which energy is transferred by all cells. However, there are other, much more specific, biosynthetic processes which are characteristic of small groups of plants. The compounds produced by this type of 'secondary' process are often completely unrelated to those produced by the secondary processes in other groups of plants.

These secondary processes involve phosphate esters of various types and they have been the subject of extensive study by organic chemists rather than by biochemists, who have concentrated on the primary processes. It is beyond the scope of this book to deal with secondary processes, and detailed descriptions can be found in texts listed in the bibliography. However, a few of the compounds produced in this manner are listed in Table 33.

5.2 Nucleic acids

The nucleic acids were so called because they were originally isolated from the cell nucleus. In 1868 Miescher, who was a student at Tübingen, became interested in the components of cell nuclei. He obtained discarded surgical dressings from the local hospital and digested the pus cells from these with a mixture of hydrochloric acid and ether. From the nuclear material which separated out of these solutions, he isolated 'nuclein', an acidic substance containing a high percentage of phosphorus; this was, in fact, the first nucleic acid ever isolated. These nuclear compounds were a continual source of interest to both chemists and early biochemists and, although they were obviously extremely complex substances, by 1930 it had become clear that there were two distinct types of nucleic acid.

HO—CH$_2$ O OH

OH OH

D-ribose
(5.21)

O
HN N
H$_2$N N N
H

guanine
(5.22)

NH$_2$
N
O N
H

cytosine
(5.23)

O
HN
O N
H

uracil
(5.24)

The first type, which was present in yeast cells, could be hydrolysed to give phosphoric acid, the carbohydrate D-ribose (5.21) and three nitrogen bases: guanine (5.22), cytosine (5.23) and uracil (5.24). The second type, which was present in the thymus of calves, could be hydrolysed to phosphoric acid, the carbohydrate deoxyribose (5.25) and four nitrogen bases: adenine (5.26), guanine (5.22), cytosine (5.23) and thymine (5.27). Because of the distinction between the carbohydrates obtained from each type, the first was called ribonucleic acid (RNA) and the second deoxyribonucleic acid (DNA). It was later discovered that RNA also gave a fourth base, adenine (5.26), on hydrolysis.

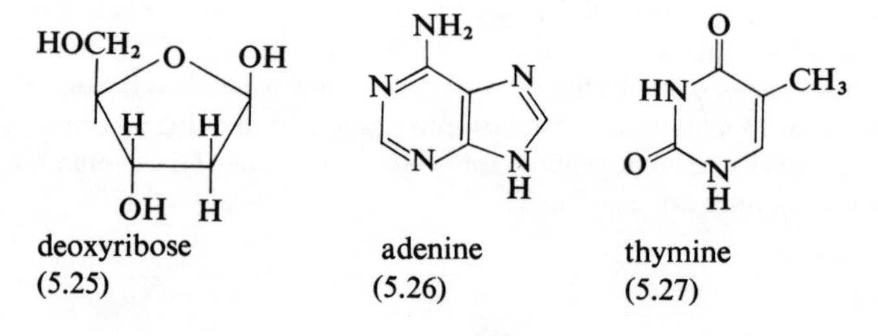

deoxyribose
(5.25)

adenine
(5.26)

thymine
(5.27)

Although attempts were made to correlate RNA with a plant origin and DNA with an animal origin, these were never very satisfactory and the only distinction in origin which really exists is that, while RNA is found throughout the cell, DNA is only found in the nucleus.

A great deal of structural information was painstakingly collected in the early part of this century, but the great leap forward in nucleic acid research came in the 1940s and 1950s when the techniques of paper and ion-exchange chromatography became generally available. These techniques enabled the chemical structure of the nucleic acids to be determined.

5.2.1 *Nucleosides*

The nucleosides are compounds which are obtained from nucleic acids by enzymic hydrolysis. Unlike the nucleic acids themselves, they do not contain phosphorus. Ribonucleic acids (RNA) yield four different nucleosides known as adenosine, guanosine, uridine and cytidine. Each of these nucleosides can be hydrolysed by acid to give a nitrogen base and a carbohydrate. Adenosine gives the base adenine (5.26), guanosine gives guanine (5.22), uridine gives uracil (5.24) and cytidine gives cytosine (5.23). The carbohydrate obtained from RNA nucleosides was shown to be D-ribose (5.21).

Deoxyribonucleic acid (DNA) also yields four nucleosides following enzymic hydrolysis; each of these can be hydrolysed by acid to give a nitrogen base and a carbohydrate. The four nitrogen bases obtained from the DNA nucleosides are adenine (5.26), guanine (5.22), cytosine (5.23) and thymine (5.27) and the carbohydrate is D-2-deoxyribose (5.25). Thus it can be seen that three nitrogen bases (adenine, guanine and cytosine) are common to both DNA and RNA, and the equivalent DNA and RNA nucleosides differ only in that one contains the carbohydrate D-2-deoxyribose while the other contains D-ribose. The fourth base, and hence the fourth nucleoside, is different in each case; RNA contains uracil and DNA contains thymine. All this information is summarized in Schemes **5.5** and **5.6**. The five nitrogen bases are derived from two skeletons, the purine skeleton (5.28) for adenine and guanine, and the pyrimidine skeleton (5.29) for uracil, cytosine and thymine.

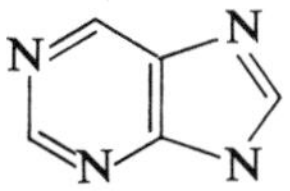

purine
(5.28)

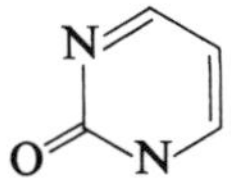

pyrimidine
(5.29)

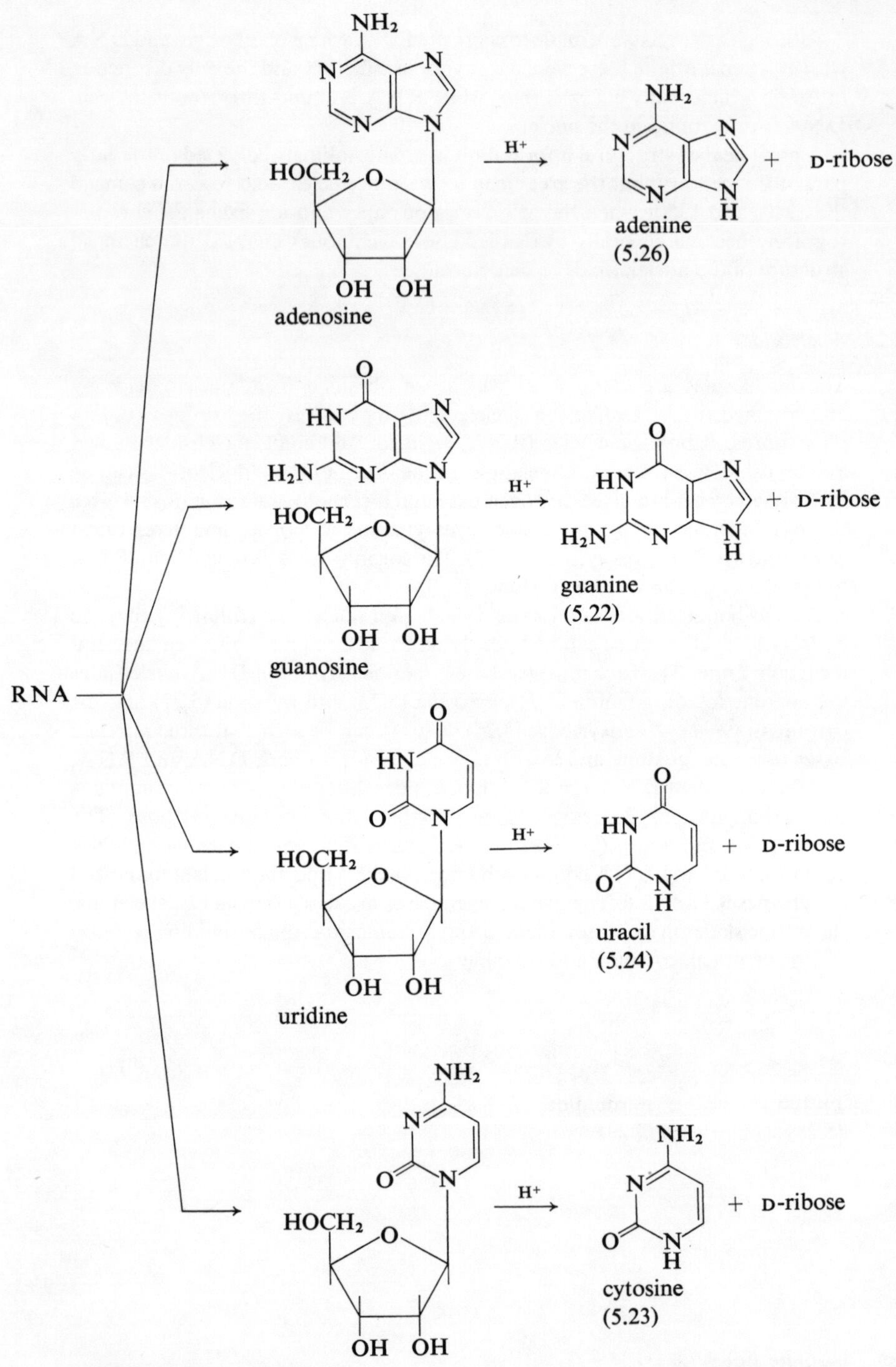

RNA
H+
adenosine
adenine
(5.26)
+ D-ribose
guanosine
guanine
(5.22)
uridine
uracil
(5.24)
cytidine
cytosine
(5.23)

Scheme **5.5**

DNA

deoxyadenosine $\xrightarrow{H^+}$ adenine (5.26) + D-2-deoxyribose

deoxyguanosine $\xrightarrow{H^+}$ guanine (5.22) + D-2-deoxyribose

deoxycytidine $\xrightarrow{H^+}$ cytosine (5.23) + D-2-deoxyribose

thymidine $\xrightarrow{H^+}$ thymine (5.27) + D-2-deoxyribose

Scheme **5.6**

If nucleosides consisted of a nitrogen base and a carbohydrate, in what positions were these linked? The carbohydrate fragment was shown to be linked, through the 1-position in each case, to the 9-position of purine (5.30) and the 3-position of pyrimidine (5.31). In the diagrams, the dash is applied to the numbering of the carbohydrate in the nucleoside to distinguish it from the numbering of the nitrogen base.

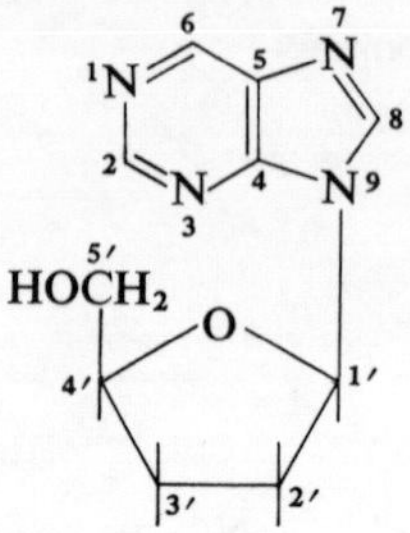

(5.30) 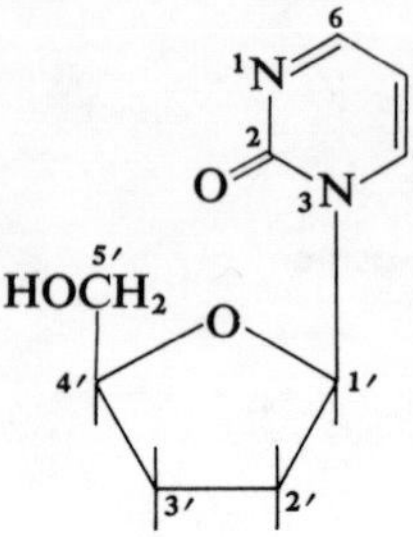 (5.31)

Elucidation of these nucleoside structures was helped by the use of ultraviolet spectroscopy and synthesis.

5.2.2 *Nucleotides*

Nucleic acids can be hydrolysed, either by specific enzymes or by very mild chemical hydrolysis, to give a mixture of nucleotides. These were quickly shown to be phosphate esters of the nucleosides that were obtained under stronger conditions of hydrolysis. In what position was the phosphate group linked to the nucleoside? Since it was shown to be linked to the carbohydrate ring, this left only three possible positions in the ribonucleotides (5.32) – 2′, 3′ or 5′ – and two in the deoxyribonucleotides (5.33) – 3′ and 5′.

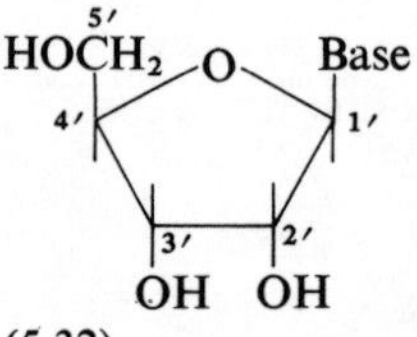

(5.32)

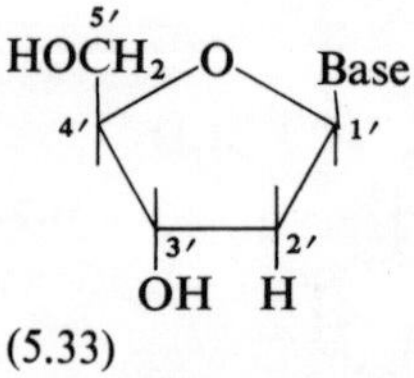

(5.33)

In fact it was found that nucleotides could be obtained with a phosphate in any of these positions, e.g. 5′-phosphate (5.34); in addition to this, a 2′,3′ cyclic phosphate (5.35) was also obtained. The isolation of the cyclic phosphate (5.35) was extremely important because it was subsequently shown that the 2′- (5.37) and 3′- (5.36) phosphates were easily interconverted, via the 2′,3′ cyclic phosphate, under acidic conditions; therefore their isolation did not specify the original position of the phosphate group in the nucleic acid.

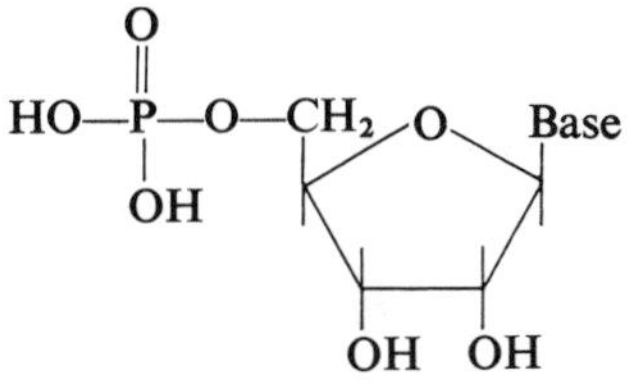

(5.34) 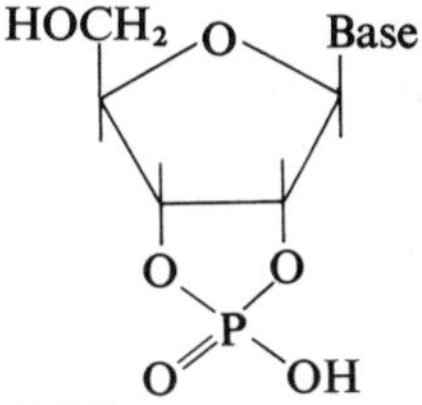 (5.35)

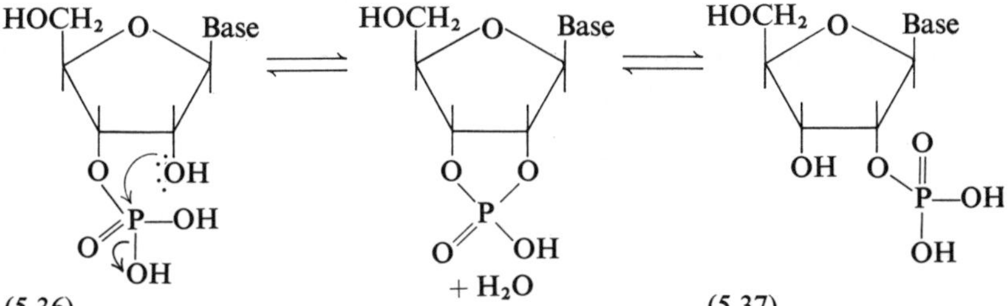

$\rightleftharpoons$ $+ H_2O$ $\rightleftharpoons$

(5.36) (5.37)

5.2.3 *The structure of RNA and DNA*

The hydrolysis products of RNA and DNA reveal that they are obviously closely related compounds, their chemical content differing mainly in the carbohydrate residue present (D-ribose for RNA and D-2′-deoxyribose for DNA). Both consist of long linear polymers of nucleosides linked by phosphate from the 5′- to the 3′-positions of the carbohydrate residues (5.38 and 5.39). The bases have not been specified because their order varies with different types of nucleic acid; the order is in fact the 'coding system' of the nucleic acids (see p. 197). The nucleosides linked in RNA are the nucleosides obtained from the hydrolysis of RNA and those linked in DNA are those obtained from its hydrolysis. However, the two nucleic acids differ enormously in molecular weight; RNA has a structure containing about a thousand nucleoside links, while DNA in its single-strand form has about two million such links.

Once the chemistry of the nucleosides and nucleotides had been worked out, the basic structure of the nucleic acids was readily deduced; the problem of their conformation remained to be solved.

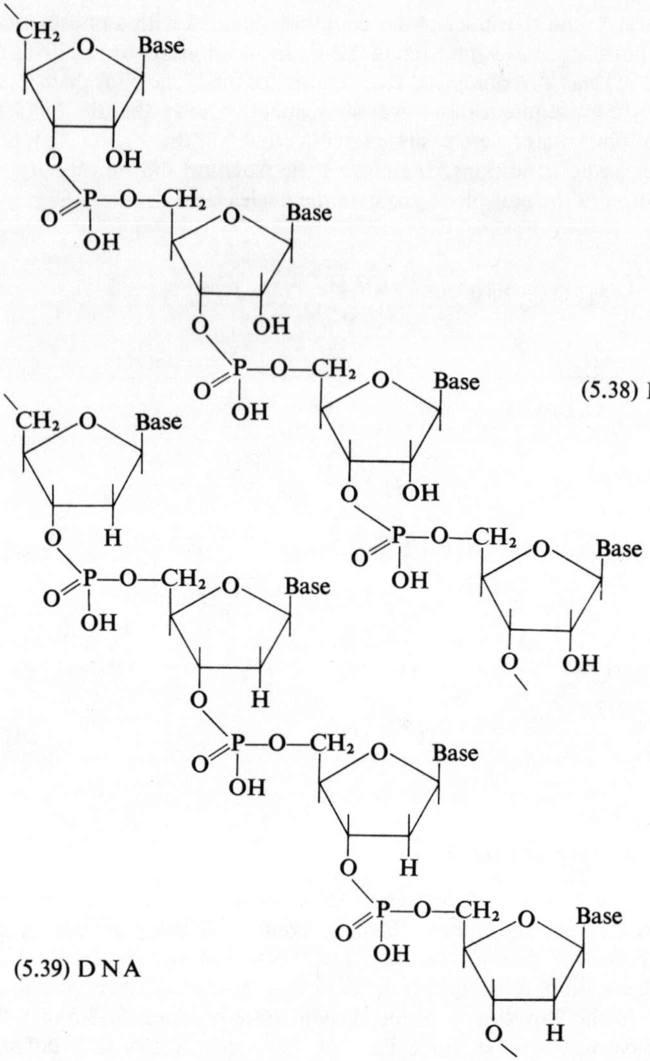

(5.38) RNA

(5.39) DNA

5.2.4 *Conformation of the nucleic acids*

By 1950, although the chemical structure of the nucleic acids was settled, little was known of their conformation. Various suggestions were put forward for the conformation of DNA, mainly based on X-ray crystallographic data, and in 1953 Pauling and Corey suggested a helical arrangement consisting of three intertwined polynucleotide chains (Figure 10). They further suggested that the polynucleotide chains were so orientated that the phosphate links were at the centre of the helix and the nitrogen bases were at the outside. It was soon pointed out that this interpretation was unacceptable on the basis of available evidence from titration experiments which appeared to show that, while the phosphorus acid groups behaved normally, the nitrogen bases were 'tied up' in some way; this was the opposite of what Pauling and Corey's structure suggested.

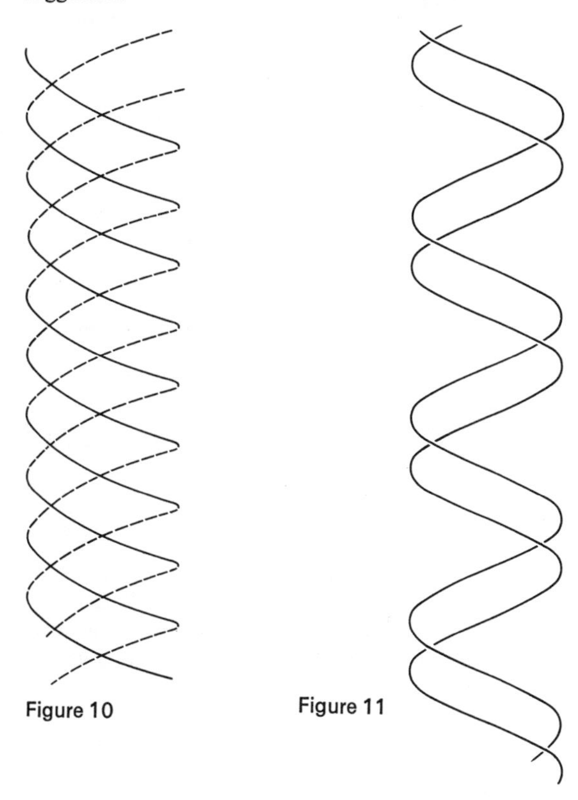

Figure 10

Figure 11

The chemical degradation of DNA had produced another piece of evidence, which proved to be vital. In the final degradation products, there appeared to be unusual regularities in the ratios of

(a) adenine: thymine,
(b) guanine: cytosine,
(c) (adenine + cytosine): (thymine + guanine).

This held true whatever the source of the DNA hydrolysed. Crick and Watson modified Pauling's theory and explained all the above phenomena.

They suggested that DNA consisted of *two* polynucleotide chains wound helically in opposite directions around a central axis (Figure 11). The titration data were readily explained if the phosphate links were at the outside of the helix, and the nitrogen bases were directed towards the central axis. However, the real genius of their theory was in the way it explained the regularities in the ratios of the bases. They postulated that the nitrogen bases in each linear chain of the helix were paired, across the axis of the helix, with the nitrogen bases in the other linear chain; in each case the pairing was made by hydrogen bonding between the bases (5.40). Since only a purine paired with a pyrimidine would conveniently fit the dimensions (which could be obtained from the X-ray measurements) across the axis of the helix, this model also explained the regular ratios of the base composition.

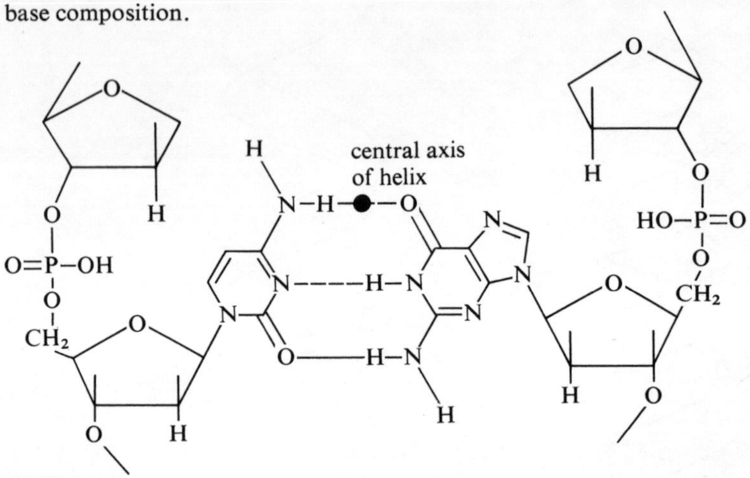

central axis
of helix

(5.40)

Detailed and sophisticated X-ray analyses carried out since 1953 have confirmed Crick and Watson's model for DNA and have shown that the arrangement of the double helix is such that a groove for a third chain is left empty (Figure 12).

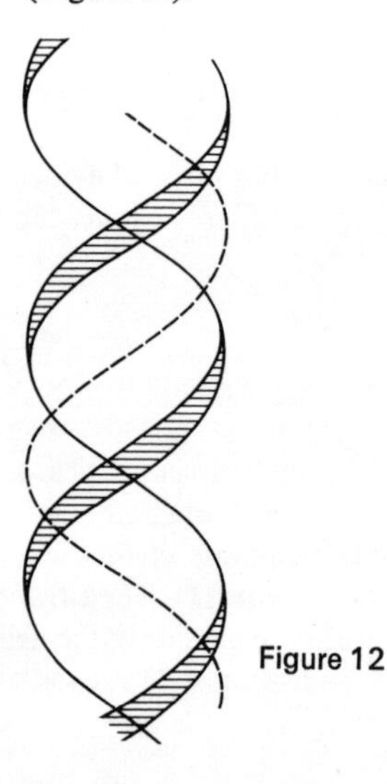

Figure 12

2.5 *Nucleic acids as carriers of information*

The nucleic acids have a vital role in the carrying and transfer of all the information a cell requires to reproduce itself. This information is detailed by the order of the mononucleotides making up the nucleic acid chain, and each piece of information is apparently determined by a small number of adjacent mononucleotides (called a *coden*).

We have already noted that DNA is found only in the cell nucleus. In fact there are probably only a very few DNA molecules in each nucleus (perhaps four or five). Molecules of most of the chemical compounds found in the cell are continually synthesized within it, but new DNA molecules are built up only at cell division.

During, or just prior to, cell division, the two strands of DNA making up the helix separate and each strand moves to one of the two new cell nuclei which have been formed from the original cell. Each of the new cells now synthesizes a second strand of DNA to exactly fit the single strand already present, and so makes a new double helix. The requirements for the synthesis of a new DNA strand are mononucleotide triphosphates (the 'building blocks'), DNA (the 'blue print') and an enzyme called DNA polymerase (the 'tools'), together with magnesium ions (Scheme **5.7**). With these components, DNA can be synthesized outside the cell in a test tube. The base-pairing that we have already discussed controls the order in which the mononucleotides are built up into the second strand of the helix and, in this way, the single strand of DNA transfers the information required for the correct formation of the second strand.

The role of RNA is at first sight more obscure, as DNA appears to carry all the information required by the cell. However, DNA molecules are unable to leave the cell nucleus and, since much of the chemical activity of the cell takes place outside the nucleus, the information from DNA must somehow be conveyed to the site of the chemical reactions. The carrier of this information is a type of RNA called *messenger* RNA. Messenger RNA is synthesized in the cell nucleus in a manner closely related to the synthesis of DNA at cell division. However, synthesis of messenger RNA takes place continuously, not just during cell divisions.

An enzyme known as RNA *polymerase* catalyses the formation of messenger RNA from its four constituent mononucleotide triphosphates. It is thought that the RNA chain is built up on a single strand of DNA (Scheme **5.8**) not on the helix structure. Since DNA molecules have a very much larger molecular weight than RNA, many RNA molecules can be built up simultaneously on the same DNA molecule. Finally, the complete RNA molecule can unwrap itself from the DNA molecule and diffuse out of the cell nucleus, carrying with it the information obtained from DNA.

deoxyadenosine triphosphate

deoxyguanosine triphosphate + single strand of DNA $\xrightarrow[\text{polymerase and } Mg^{2+}]{\text{DNA}}$ double helix of DNA + HOPOPOH

deoxycytidine triphosphate

thymidine triphosphate

Scheme **5.7**

adenosine triphosphate

guanosine triphosphate

uridine triphosphate

cytidine triphosphate

+ single strand of DNA $\xrightarrow[\text{polymerase}]{\text{RNA}}$

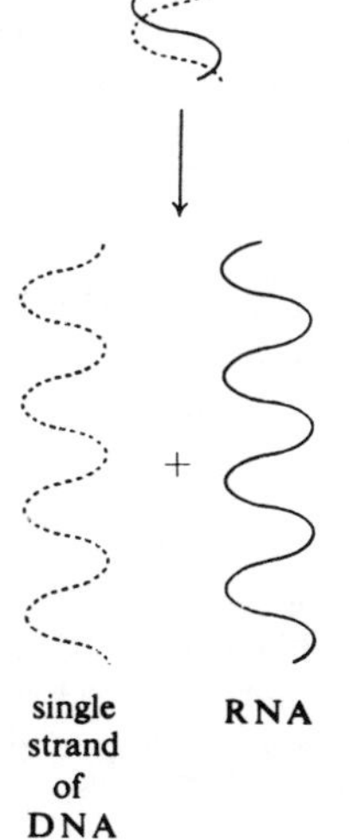

Scheme **5.8**

5.2.6 *Protein synthesis by RNA*

The messenger RNA, after leaving the cell nucleus, attaches itself to cell components known as ribosomes. Other, shorter-chained, RNA molecules known as *transfer* RNA then link up with the messenger RNA by specific base-pairing. Each amino acid required by the cell to synthesize proteins has a specific, different, transfer RNA molecule to which it can attach itself. When these transfer RNA molecules, each with an amino acid attached, have joined the messenger RNA, the amino acids are in the correct position to be built into a protein chain on the surface of the ribosome. It is the specific base-pairing of the transfer RNA with the messenger RNA chain which controls the order of amino acids in the protein chain. As we have seen, this information derives originally from the DNA in the cell nucleus.

DNA, and hence messenger RNA, contains the information necessary for the synthesis of all cell components, of which proteins are but an example. In addition to this, DNA also contains the information necessary to assemble all these component molecules into a complete functioning cell. The total quantity of this information is incredibly large, and we are only just beginning to appreciate the full significance of DNA molecules.

5.3 Organophosphorus insecticides and nerve gases

The first indications of insecticidal activity among organophosphorus compounds were found in 1930, but the first compound of this type, HETP (an impure mixture containing tetraethyl pyrophosphate (5.41) as the active ingredient) was not used as an agricultural insecticide until 1942. Extra impetus was undoubtedly given to this type of research by World War II, and the potential use of these compounds as gas weapons. Manufacturing units capable of producing several hundred tons of nerve gas per month were discovered in Germany after the war, and it would be naïve to think that the Allies were not similarly prepared. Fortunately, good sense has so far prevailed and these compounds have never been used as a weapon of war, although there are a number of examples of their use in espionage.

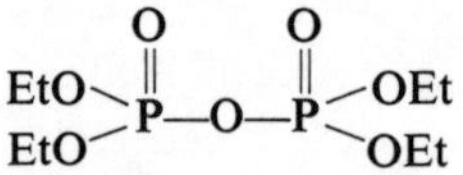

$$(EtO)_2P(=O)\text{—O—}P(=O)(OEt)_2$$

(5.41)

Since the first introduction of HETP, the number of organophosphorus insecticides available has risen to several hundred (for examples see Table 34). HETP itself is fairly readily hydrolysed, so its effects are short-term, and more-stable, longer-lasting compounds (e.g. Parathion; 5.42) have tended to replace HETP. Even more fundamental was the discovery that certain compounds (e.g. Schradan; 5.43) were absorbed by the plant and so were active over the whole organism, even when only one part had been initially treated. These

Table 34 Organophosphorus insecticides

Insecticide	*Preparation*	*Mammalian toxicity (rat)* LD_{50}/ mg kg^{-1})	*Insecticidal toxicity (typical values)* LD_{50}/(mg kg^{-1})
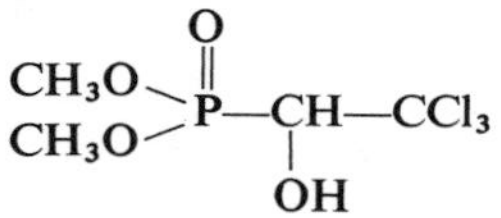 $(CH_3O)_2P(=O)-CH(OH)-CCl_3$ Dipterex *O,O*-dimethyl (1-hydroxy-2,2,2-trichloroethyl)-phosphonate	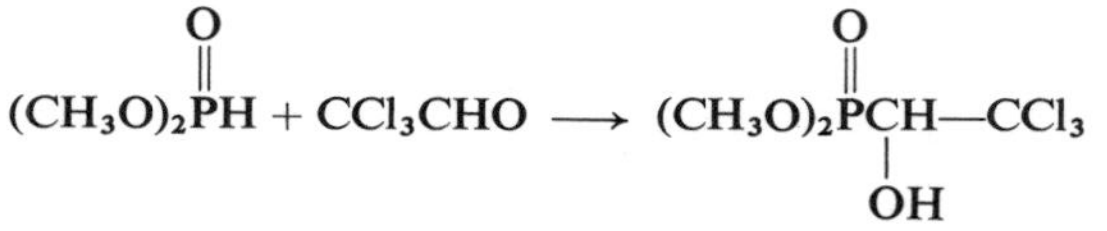 $(CH_3O)_2P(=O)H + CCl_3CHO \longrightarrow (CH_3O)_2P(=O)CH(OH)-CCl_3$ b.p. 120 °C at 0·4 mm	600	30
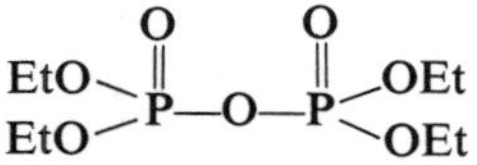 $(EtO)_2P(=O)-O-P(=O)(OEt)_2$ TEPP tetraethyl pyrophosphate (This was the main constituent of HETP)	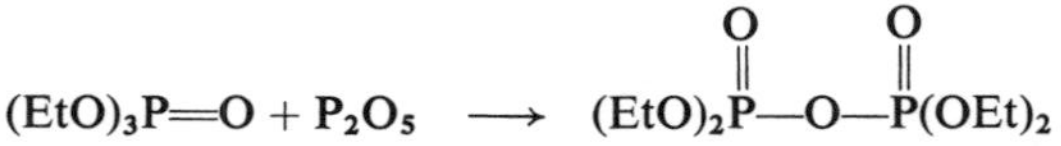 $(EtO)_3P{=}O + P_2O_5 \longrightarrow (EtO)_2P(=O)-O-P(=O)(OEt)_2$ b.p. 144 °C at 3 mm	1·12	15
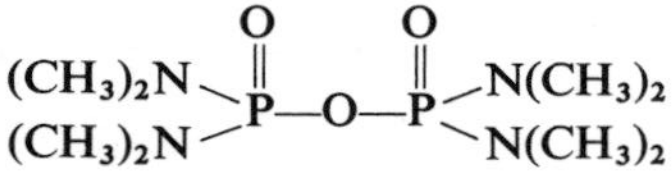 $[(CH_3)_2N]_2P(=O)-O-P(=O)[N(CH_3)_2]_2$ Schradan bis(dimethylamino)-phosphoric anhydride	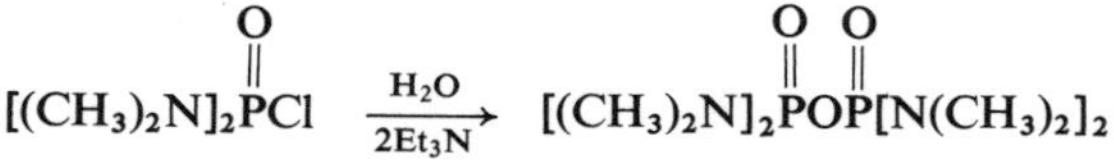 $[(CH_3)_2N]_2P(=O)Cl \xrightarrow[2Et_3N]{H_2O} [(CH_3)_2N]_2P(=O)OP(=O)[N(CH_3)_2]_2$ b.p. 142 °C at 2 mm	20	120

Table 34 – *continued*

Insecticide	*Preparation*	*Mammalian toxicity (rat)* LD_{50}/(mg kg^{-1})	*Insecticidal toxicity (typical values)* LD_{50}/(mg kg^{-1})
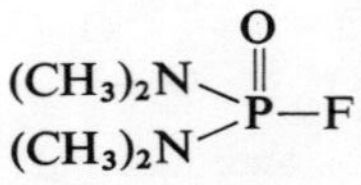 $[(CH_3)_2N]_2P(=O)F$ Dimefox bis(dimethylamino)-phosphoryl fluoride	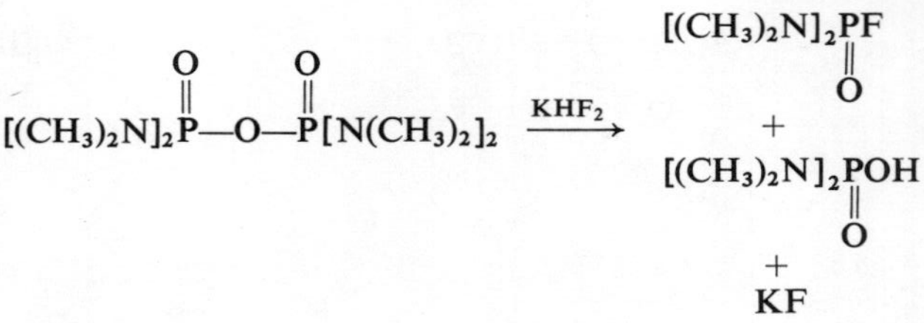 $[(CH_3)_2N]_2P(=O){-}O{-}P(=O)[N(CH_3)_2]_2 \xrightarrow{KHF_2} [(CH_3)_2N]_2P(=O)F + [(CH_3)_2N]_2P(=O)OH + KF$	7	
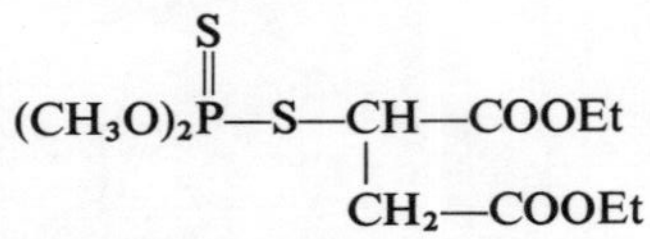 $(CH_3O)_2P(=S){-}S{-}CH(COOEt){-}CH_2{-}COOEt$ Malathion *O*,*O*-dimethyl *S*-(1,2-dicarboxyethyl) dithiophosphate	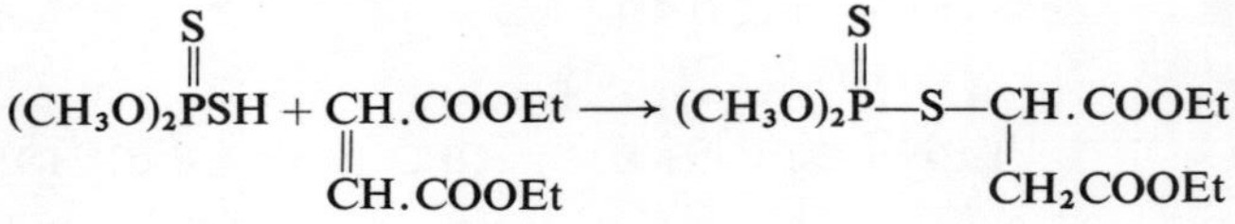 $(CH_3O)_2P(=S)SH + CH.COOEt{=}CH.COOEt \longrightarrow (CH_3O)_2P(=S){-}S{-}CH(COOEt){-}CH_2COOEt$ b.p. 146°C at 0·8 mm	100	

$(CH_3)_2CH$–(pyrimidine, 4-CH_3)–O–P(=S)$(OEt)_2$ Diazinon *O*,*O*-diethyl *O*-(2-isopropyl- 4-methylpyrimidyl-6) thiophosphate	$(CH_3)_2CH$–(pyrimidine, 4-CH_3)–OH + $(EtO)_2P(=S)$—Cl ↓ $(CH_3)_2CH$–(pyrimidine, 4-CH_3)–O—P(=S)$(OEt)_2$ b.p. 125 °C at 1 mm	100	240
$(EtO)_2P(=S)$—O–C_6H_4–NO_2 Parathion *O*,*O*-diethyl *O*-(*p*-nitrophenyl) thiophosphate	$(EtO)_2P(=S)Cl$ + HO–C_6H_4–NO_2 ↓ $(EtO)_2P(=S)$—O–C_6H_4–NO_2	6	3·5

Table 34 – *continued*

Insecticide	*Preparation*	*Mammalian toxicity (rat)* $LD_{50}/(mg\,kg^{-1})$	*Insecticidal toxicity (typical values)* $LD_{50}/(mg\,kg^{-1})$
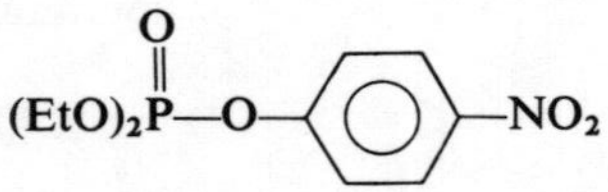 $(EtO)_2P(=O)—O—C_6H_4—NO_2$ Paraoxon *O,O*-diethyl *O*-(*p*-nitrophenyl) phosphate	similar to Parathion using $(EtO)_2P(=O)—Cl$ b.p. 149 °C at 0·8 mm	3	0·7
$(EtO)_2P(=O)—S—CH_2CH_2SEt$ and $(EtO)_2P(=S)—O—CH_2CH_2SEt$ Systox *O,O*-diethyl *O*-2-(ethylmercapto)ethyl thiophosphate	$(EtO)_3P + HOCH_2CH_2SEt$ ↓ $(EtO)_2P—O—CH_2CH_2SEt$ ↓ Systox b.p. 138 °C at 2·5 mm	1–30	

$$(EtO)_2P(=S)-O-C_6H_4-NO_2$$

(5.42)

$$(Me_2N)_2P(=O)-O-P(=O)(NMe_2)_2$$

(5.43)

compounds are known as *systemic* insecticides, since they become part of the plant's system.

Unfortunately, the early insecticides were also extremely toxic to mammals (see Table 34). This led to problems involving both the effect of residues on crops when they were harvested, and on the personnel employed in application. The first problem was readily overcome by legislation which controlled the minimum time between spraying and harvesting; this is not usually a serious problem anyway, since most compounds are quickly broken down. However, the second problem has prompted much investigation, aimed at reducing mammalian toxicity while maintaining effective insecticidal properties. This is not an easy problem to solve, since the mechanisms of poisoning in each case are closely related. However, insecticides with quite low mammalian toxicity are now known (e.g. Malathion; 5.44 and Ekatin; 5.45). The figures for toxicity are given in Table 34, and 'low' is, of course, a relative term. Toxicity is usually measured in terms of LD_{50}, which is the average minimum dosage, in milligrammes per kilogramme of body weight, which will kill 50 per cent of a group of a particular species of animal. Malathion, for example, has an oral LD_{50} for rats of about 1500 mg kg^{-1}, while the value for Ekatin is 225 mg kg^{-1}. Thus, if the same LD_{50} values also apply to other species (in fact they often vary markedly), in a group of men each weighing seventy-five kilogrammes (twelve stone), 120 g of Malathion, or alternatively 20 g of Ekatin, would be required by each man in order to kill 50 per cent of them. On the other hand HETP has an LD_{50} for mice of 0·65 mg kg^{-1} (52 mg would kill 50 per cent of the same men) and compounds undoubtedly exist that have LD_{50}s lower than 0·01 mg kg^{-1} (less than one milligramme would kill 50 per cent of the men).

$$(CH_3O)_2P(=S)-S-CH(COOC_2H_5)-CH_2-COOC_2H_5$$

(5.44)

$$(CH_3-O)_2P(=S)-S-CH_2-CH_2-S-CH_2-CH_3$$

(5.45)

Several species of insects have developed resistance to organophosphorus insecticides, as they did to chlorinated hydrocarbons. As might be expected, since resistance appears to develop through a process of selection, resistance has appeared in those species which have been particularly intensely and effectively attacked (e.g. mosquitos). However, a phenomenon known as 'potentiation' has been observed in which a mixture of two insecticides will sometimes have a much higher potency than the same two acting separately.

Another type of potentiation effect is known, whereby compounds which do not appear to be toxic *in vitro* are converted into much more toxic compounds by the metabolism of the insect, or animal. An example of this type is Parathion (5.42), which is a poor phosphorylating agent *in vitro* but *in vivo* is rapidly oxidized to Paraoxon (5.46), which is extremely toxic.

$$(EtO)_2P(=S)-O-C_6H_4-NO_2 \xrightarrow{[O]} (EtO)_2P(=O)-O-C_6H_4-NO_2$$

(5.42) (5.46)

$$R_2P(=X)Y$$

(5.47)

A cross-section of active insecticides, together with their methods of preparation and LD_{50} values, is given in Table 34. The general structure required for activity can be summarized as (5.47), where R are alkoxy or amino groups, X is oxygen or sulphur and Y is a good anionic leaving group. We shall see later that it is displacement of Y by the active site of an enzyme (e.g. acetylcholinesterase) which accounts for the toxic effects of these compounds.

5.3.1 *Role of acetylcholinesterase in nerve function*

(a) *The mammalian nervous system.* Overwhelming evidence is now available that organophosphorus compounds with 'nerve gas' properties act by inhibiting the action of enzymes involved in nerve function. Most of their effects appear to be due to the inhibition of one particular enzyme, acetylcholinesterase. In order to understand the nature of this inhibition, we need to know the part that acetylcholinesterase plays in nerve functions in mammals. Although a full description of nerve function would require far more space than is available here, a much-simplified picture will serve our present purposes.

Nerve fibres have an electrical potential of about 100 mV across their outer wall (Figure 13) and, when a nerve is stimulated at one end (e.g. by a message from the brain), the sign of this potential is momentarily reversed over a small area of the nerve fibre (Figure 14). This small area of reversed potential travels rapidly (about 122 m s^{-1} or 400 ft s^{-1}) along the nerve fibre until it reaches the nerve ending, where it stimulates the release of acetylcholine (5.48).

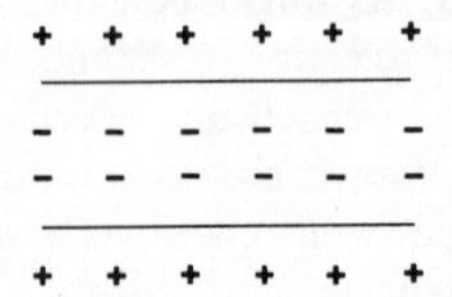

Figure 13 Unstimulated nerve fibre

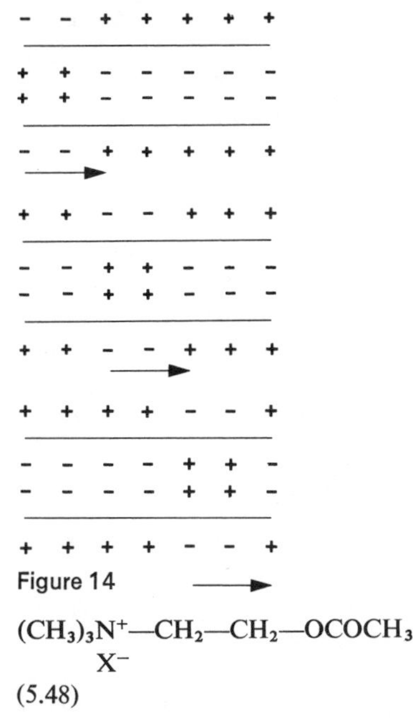

Figure 14

$$(CH_3)_3N^+—CH_2—CH_2—OCOCH_3$$
$$X^-$$

(5.48)

Let us consider a nerve which is supplying a muscle with impulses, although the principles that we shall discuss will apply equally well to nerves supplying other sites like glands or other nerve routes. The nerve ending from which acetylcholine is released is not directly connected to the surface of the muscle that it is supplying; there is a gap between the two of about 10^{-8} m (Figure 15).

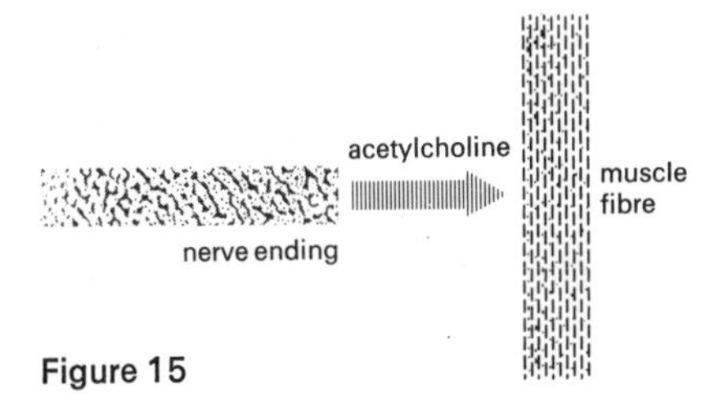

Figure 15

The acetylcholine produced by stimulation of the nerve diffuses across this gap until it reaches the muscle surface, where it causes a change in electrical potential and initiates a contraction. On arriving at the muscle surface, the acetylcholine is very rapidly hydrolysed by an enzyme, acetylcholinesterase (*esterases* are enzymes that hydrolyse ester linkages). This stops the stimulation and allows the muscle to relax. Before another contraction can take place, a further nerve stimulation and acetylcholine release is required. However, a normal muscle–nerve system can take part in perhaps a thousand of these processes – involving stimulation, contraction and relaxation – every second.

Organophosphorus nerve poisons (or anti-cholinesterases) are now known to interfere with this process of acetylcholine hydrolysis by deactivating the enzyme responsible (acetylcholinesterase, see p. 206). Since its hydrolysis is inhibited, the acetylcholine can only leave the area of the muscle by diffusion, which is a comparatively slow process. Because of this, a second stimulation, and consequent release of acetylcholine, will arrive while the original acetylcholine is still present at the muscle surface. This leads to complete confusion of cause and effect, and the muscle being stimulated undergoes continuous contraction (tetanus). If one imagines this process taking place at every nerve junction at every muscle, gland and organ in the body, together with other more obscure effects caused by diffusion into the central nervous system, something of its significance can be appreciated!

Symptoms of poisoning in this way are rapid twitching (fasciculation) of muscles, followed by paralysis, intense pupil contraction, excessive saliva production and more complex effects, such as loss of coordination, caused by fundamental blockages in the central nervous system.

On the other hand, these same organophosphorus anti-cholinesterases can also be used in the treatment of certain diseases. Glaucoma is a disease which causes an increase in pressure in the fluid inside the eye and, if untreated, may cause irreversible damage to the optic nerve. Anti-cholinesterases reduce this pressure and have been used to some extent in treatment, although they appear to cause unpleasant side effects in some patients. Some success has also been obtained in treatment of myasthenia gravis, the symptoms of which are the rapid fatigue of muscles. The cause of this is a very low acetylcholine concentration at the nerve junctions and so an acetylcholinesterase inhibitor should be therapeutic.

(b) *The insect nervous system.* The nervous system of insects has been much less studied than that of mammals and consequently is not so well understood. Several differences between the two types have been found, but insects appear to contain fairly large amounts of acetylcholinesterase and acetylcholine. However, it is not clear whether these substances play the same vital role that they do in mammals; in fact some evidence points to the ability of certain insects to remain healthy with all their acetylcholinesterase inhibited (this would cause death in mammals). Equally, some insects killed by pesticide treatment have been found to have much of their acetylcholinesterase uninhibited.

Probably the best explanation of these facts is that, while enzyme inhibition is still responsible for the toxic effects of cholinergic substances in insects, the vital enzymes inhibited are different from those in mammals.

5.3.2 *Mechanism of inhibition of acetylcholinesterase*

Two questions arise from the preceding discussion of the role of acetylcholinesterase. Firstly, what evidence exists that it is inhibition of cholinesterase, rather than some other enzyme, which causes organophosphorus poisoning in mammals? Secondly, what is the specific mechanism of the inhibition?

Evidence in favour of the vital role of cholinesterase inhibition is considerable. Acetylcholinesterase appears to be the only enzyme which shows a good correlation between the extent of inhibition and the effect of poisoning over a large range of different inhibitors. The duration of poisoning symptoms agrees with the known lifetime of phosphorylated cholinesterase (see p. 210) *in vivo*. Finally, compounds which specifically reactivate inhibited cholinesterase also reduce toxic symptoms.

As has already been suggested, the inhibition of cholinesterase by certain organophosphorus compounds is caused by phosphorylation of the enzyme. In terms of their interaction with phosphorus compounds, esterases can be divided into three groups:

(a) those that do not interact;
(b) those that can hydrolyse phosphate esters;
(c) those that are inhibited by certain phosphate esters.

Obviously we need not consider those of type (a). Let us discuss those of type (c) first since, after this, the significance of type (b) will become apparent.

The mechanism of hydrolysis of acetylcholine by cholinesterase can be simply represented by the initial reversible formation of a complex (5.49) between the enzyme and acetylcholine (5.48), followed by transfer of the acetyl group of choline to the enzyme, giving an acetylated enzyme (5.50) and free choline (5.51). Finally the esterase is hydrolysed by water to give acetic acid and free enzyme, which can then hydrolyse more acetylcholine.

$(CH_3)_3N^+—CH_2—CH_2—OCOCH_3$ + acetylcholinesterase
X^-

(5.48)

$\rightleftharpoons$

complex of acetylcholine absorbed on enzyme surface

(5.49)

$\downarrow$

CH_3CO—enzyme + $(CH_3)_3N^+—CH_2CH_2—OH$
X^-

(5.50) (5.51)

$\downarrow H_2O$

CH_3COOH + free acetylcholinesterase

Acetylcholinesterase will carry out a very similar reaction with any compound capable of transferring a phosphoryl group; this is analogous to the transfer of an acetyl group by acetylcholine. A similar initial complex (5.51) is formed,

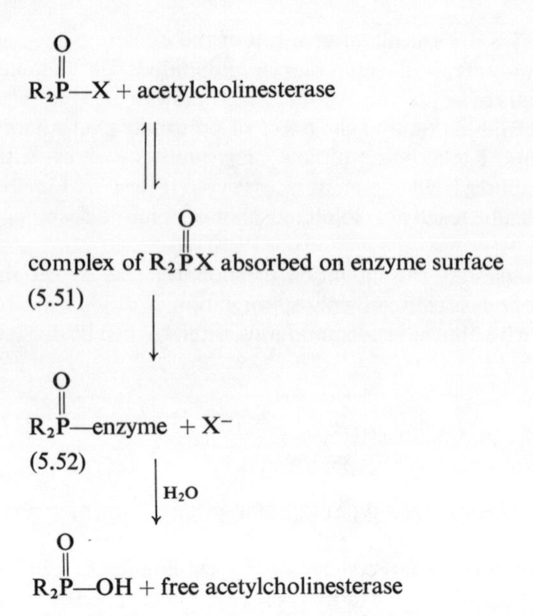

followed by transfer of a phosphoryl group to the enzyme to give phosphorylated cholinesterase (5.52). Finally, hydrolysis can take place to give a phosphorus acid and free acetylcholinesterase.

Whether any particular enzyme will be inhibited by this type of process, that is, whether it is type (c), depends on the comparative rate of hydrolysis back to free enzyme and phosphorus acid; since, while it is phosphorylated at its active site, it cannot carry out its true function of hydrolysing other esters (e.g. acetylcholine).

Phosphorylated acetylcholinesterase is very slowly hydrolysed *in vivo*, so acetylcholinesterase is largely inhibited by this phosphorylation. Reactivation could in theory be achieved if the rate of hydrolysis of the phosphorylated enzyme could be increased, but conditions *in vivo* are not easily modified. Some success has been achieved in this direction by the use of oximes.

The pyridinium oxime (5.54) has been used to treat cases of nerve poisoning. It is thought that some interaction involving the N^+ ion holds the oxime in the correct position on the enzyme to hydrolyse the phosphate link.

$$I^- \; N^+(CH_3)\text{-pyridine-}CH{=}N{-}OH + R_2P(=O){-}\text{cholinesterase} \longrightarrow I^- \; N^+(CH_3)\text{-pyridine-}CH{=}N{-}O{-}P(=O)R_2 + \text{free cholinesterase}$$

(5.54)

Chapter 6
Radical and Related Reactions of Phosphorus Compounds

.1 Phosphorus free radicals

Apart from their purely chemical interest, phosphorus radicals are important biologically. For example, Parathion (6.1), a widely used insecticide, has only a

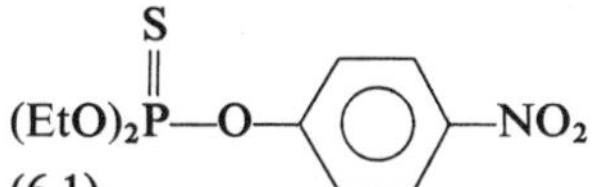

$$(EtO)_2\overset{\overset{S}{\|}}{P}-O-C_6H_4-NO_2$$

(6.1)

moderate cholinesterase-inhibiting effect in mammals (see p. 208), and is therefore apparently safe to use on crops. However, ultraviolet irradiation of Parathion leads to a large increase in its toxicity to mammals, an increase which could conceivably be accomplished by sunlight after spraying. The chemistry of this process is not fully understood but may well involve phosphorus radicals. Fortunately, the rate of hydrolysis of such compounds in the soil is usually high and they are quickly deactivated.

It is important to understand the effect of high-energy radiation on phosphorus compounds, since ionizing radiation (X-rays or γ-rays) is being increasingly used in food preservation. Phosphate esters are also extensively used in complexing out uranium from reactor fuels.

For these, and other reasons, there has been a steadily increasing interest in the chemistry of phosphorus radicals. This differs considerably from the chemistry of nitrogen radicals, largely because phosphorus is able to expand its outer electron shell beyond an octet.

There are two distinct types of phosphorus radical, although both may be involved in any particular radical reaction. Those which are analogous to nitrogen radicals (6.2) are known as phosphinyl radicals (6.3) and have seven outer electrons, all of which can be accommodated in the 3s and 3p orbitals. There are also phosphorus radicals which have no nitrogen analogues. These are known as phosphoranyl radicals (6.4) and, having nine outer electrons, must populate the 3d orbital with at least one electron. Radical cations (6.5)

$R_2\ddot{N}\cdot$	$R_2\ddot{P}\cdot$	$R_4\dot{P}$	$[R_3P\cdot]^+$	$[R_3\ddot{P}\cdot]^-$
(6.2)	(6.3)	(6.4)	(6.5)	(6.6)

(seven outer electrons) and radical anions (6.6) (nine outer electrons) are also possible.

Phosphinyl radicals are prepared by homolytic fission of a phosphorus–X bond; either energetically, by heat or radiation (Scheme **6.1**) or by radical displacement at X (Scheme **6.2**).

$$R_2\ddot{P}—\ddot{P}R_2 \xrightarrow[\text{or } h\nu]{\text{heat}} 2R_2\ddot{P}\cdot$$

(6.7)

Scheme **6.1**

$$R_2\ddot{P}—H + R\cdot' \longrightarrow R_2\ddot{P}\cdot + R'H$$

Scheme **6.2**

Phosphinyl radicals are generally fairly stable, and hence tend to be specific in their reactions. Diphenylphosphinyl radicals, prepared by low-temperature ultraviolet irradiation of tetraphenyldiphosphine (6.7; R = Ph), have a half-life of some twenty minutes at 173 K and apparently an indefinite lifetime at 77 K. They normally undergo one of two types of reaction, addition (Scheme **6.3**) or abstraction (Scheme **6.4**).

$$R_2\ddot{P}\cdot + Y{=}X \longrightarrow R_2\ddot{P}—Y—X\cdot$$

Scheme **6.3**

$$R_2\ddot{P}\cdot + Y—X \longrightarrow R_2\ddot{P}—Y + X\cdot$$

Scheme **6.4**

Phosphoranyl radicals are prepared either by radical addition to phosphorus(III) (Scheme **6.5**), or by a one-electron reduction of phosphonium salts (Scheme **6.6**).

$$R_3P{:} + R\cdot' \longrightarrow R_3\dot{P}—R'$$

Scheme **6.5**

$$R_4P^+ + e \longrightarrow R_4P\cdot$$

Scheme **6.6**

Reactions involving phosphoranyl radicals are generally much more complex than those involving the more stable phosphinyl radicals. Phosphoranyl radicals undergo two types of reaction: α-scission (Scheme **6.7**), which is the reverse of one method of formation, or β-scission (Scheme **6.8**), which usually involves P=O bond formation. The reactions involving phosphoranyl radicals are those generally thought of as unique phosphorus radical reactions.

$$R_3\dot{P}—R' \longrightarrow R_3P{:} + \cdot R'$$

Scheme **6.7**

$$R_3\dot{P}—O—R' \longrightarrow R_3P{=}O + \cdot R'$$

Scheme **6.8**

Phosphonyl radicals (6.8) are also known, and could be thought of as phosphinyl radicals with the lone pair coordinated to oxygen or, if the d-orbital contribution to P—O bonding is considered, as phosphoranyl radicals. However,

$$R_2\overset{\overset{\displaystyle O}{\|}}{P}\cdot$$

(6.8)

the unpaired electron appears to be little affected by the involvement of d-orbitals, and the reactions of phosphonyl radicals are closely analogous to those of phosphinyl radicals.

1.1 *Phosphinyl radicals*

(a) *Abstraction reactions.* In the absence of alternative pathways, phosphinyl radicals will often abstract hydrogen, or some other atom, from the reaction solvent. Photolysis of triphenylphosphine in benzene under a nitrogen atmosphere leads to the formation of biphenyl (6.9), diphenylphosphine (6.10) and tetraphenyldiphosphine (6.11), presumably by the initial formation of diphenylphosphinyl and phenyl radicals followed by both abstraction of hydrogen from the solvent and dimerization.

$$Ph_3P: \longrightarrow Ph_2\ddot{P}\cdot + Ph\cdot$$

$$Ph_2\ddot{P}\cdot + PhH \longrightarrow \underset{(6.10)}{Ph_2\ddot{P}H} + Ph\cdot$$

$$2Ph_2\ddot{P}\cdot \longrightarrow \underset{(6.11)}{Ph_2P\text{—}PPh_2}$$

$$2Ph\cdot \longrightarrow \underset{(6.9)}{Ph\text{—}Ph}$$

$$\underset{(6.11)}{Ph_2P\text{—}PPh_2} \xrightarrow{h\nu} 2Ph_2\ddot{P}\cdot$$

$$Ph_2\ddot{P}\cdot + H\text{—}\ddot{\underset{..}{O}}\text{—}R \longrightarrow Ph_2\dot{\ddot{P}}{}^{-}\text{—}\underset{\substack{|\\H}}{\ddot{O}{}^{+}}\text{—}R \xrightarrow{Ph_2\ddot{P}\cdot} Ph_2\ddot{P}\text{—}O\text{—}R + \underset{(6.10)}{Ph_2\ddot{P}H}$$

$$Ph_2\ddot{P}\text{—}O\text{—}R \longrightarrow \underset{(6.12)}{Ph_2\overset{\overset{O}{\|}}{P}\text{—}R}$$

However, some apparently simple radical-abstraction reactions are more complex than the products suggest. For example, tetraphenyldiphosphine (6.11), on photolysis (or thermolysis) in the presence of alcohols, forms diphenylphosphine (6.10) and a phosphine oxide (6.12). The reaction is thought to proceed by the attack of a diphenylphosphinyl radical on an alcohol oxygen atom rather than by simple hydrogen abstraction. This is because carbon radicals behave quite differently in their reaction with alcohols, abstracting a hydrogen atom linked to a carbon rather than to an oxygen atom. This difference in behaviour is difficult to explain unless fundamental differences between phosphorus and carbon (for example, the ability of phosphorus to expand its octet, or its extreme affinity for oxygen) are involved in the reaction pathway.

$$R_3C\cdot + R'_2CH{-}O{-}H \longrightarrow R_3CH + R'_2\dot{C}{-}O{-}H$$

(b) *Addition reactions.* The addition to carbon–carbon and carbon–oxygen double bonds is the best-known reaction of phosphinyl radicals (see Table 35).

Phosphine (PH_3) will add to olefins in the presence of chemical (or photochemical) initiation to give mixtures of primary, secondary and tertiary phosphines by the mechanism outlined, with phosphorus adding primarily to the least-substituted end of the double bond. That this type of addition is reversible is shown by the ability of traces of phosphinyl radicals to isomerize *cis*-olefins to *trans*-olefins.

$$PH_3 \xrightarrow[h\nu]{(PhCOO)_2} H_2P\cdot$$

$$H_2P\cdot + R_2C{=}CR_2 \longrightarrow H_2P{-}CR_2{-}\dot{C}R_2$$

$$H_2P{-}CR_2{-}\dot{C}R_2 + PH_3 \longrightarrow H_2P{-}CR_2{-}CHR_2 + \dot{P}H_2$$

$$H_2P{-}CR_2CR_2H \longrightarrow H\dot{P}{-}CR_2{-}CR_2H$$

$$H\dot{P}{-}CR_2{-}CR_2H + R_2C{=}CR_2 \longrightarrow HP(CR_2{-}CHR_2){-}CR_2{-}\dot{C}R_2$$

$$\underset{\textit{cis}\text{-2-butene}}{(H)(CH_3)C{=}C(H)(CH_3)} \xrightarrow{R_2P\cdot} \underset{\textit{trans}\text{-2-butene}}{(CH_3)(H)C{=}C(H)(CH_3)}$$

Dialkyl phosphites (6.13) and phosphonous acids (6.14) undergo analogous additions by a similar mechanism. The dialkyl phosphite reaction is very useful synthetically because of its high yields (50–90 per cent) and mild conditions, and it has been widely used commercially. Similar additions to carbonyl double bonds take place to give α-hydroxyphosphonates (6.15).

$$\underset{(6.13)}{(RO)_2P(=O)H} + R{-}CH{=}CH_2 \xrightarrow{\text{initiator}} (RO)_2P(=O)CH_2CH_2R$$

$$\underset{(6.14)}{H_2P(=O){-}OH} + R{-}CH{=}CH_2 \xrightarrow{\text{initiator}} H{-}P(=O)(OH){-}CH_2{-}CH_2{-}R$$

Table 35 Radical addition reactions

Phosphines
$PH_3 + 3CH_3(CH_2)_5{-}CH{=}CH_2 \xrightarrow[50\ \text{p.s.i.}]{100\,°C} (CH_3(CH_2)_5{-}CH_2{-}CH_2)_3P$ 83%
$4PH_3 + CH_3(CH_2)_5{-}CH{=}CH_2 \xrightarrow[500\ \text{p.s.i.}]{80\,°C} (CH_3(CH_2)_5{-}CH_2{-}CH_2)_3P$ 4% + $(CH_3(CH_2)_5{-}CH_2{-}CH_2)_2PH$ 18% + $(CH_3(CH_2)_5{-}CH_2{-}CH_2)PH_2$ 65%
$NC{-}CH_2CH_2PH_2 + 2CH_3(CH_2)_5CH{=}CH_2 \xrightarrow{80\,°C} NCCH_2CH_2P(CH_2(CH_2)_6CH_3)_2$ 93%
$NC{-}CH_2CH_2PH_2$ + cyclohexene $\xrightarrow{120\,°C} NC{-}CH_2{-}CH_2\overset{H}{P}{-}C_6H_{11}$ (cyclohexyl) 65%
Phosphites
$(CH_3O)_2\overset{O}{\overset{\parallel}{P}}H + 3CH_2{=}CH{-}(CH_2)_8\overset{O}{\overset{\parallel}{C}}OCH_3 \xrightarrow[110\,°C]{100-} (CH_3O)_2\overset{O}{\overset{\parallel}{P}}(CH_2)_{10}\overset{O}{\overset{\parallel}{C}}OCH_3$ (60%)
$(CH_3O)_2\overset{O}{\overset{\parallel}{P}}H + 3CH_3(CH_2)_7CH{=}CH(CH_2)_7\overset{O}{\overset{\parallel}{C}}OCH_3 \xrightarrow{100\,°C}$ $CH_3(CH_2)_7\underset{(CH_3O)_2P{=}O}{\underset{\mid}{CH}}.(CH_2)_8.\overset{O}{\overset{\parallel}{C}}{-}OCH_3 + CH_3(CH_2)_8\underset{(CH_3O)_2P{=}O}{\underset{\mid}{CH}}{-}(CH_2)_7\overset{O}{\overset{\parallel}{C}}{-}OCH_3$ 68% total yield

$$(RO)_2\overset{O}{\overset{\parallel}{P}}H + CH_3.CO.CH_3 \longrightarrow (RO)_2\overset{O}{\overset{\parallel}{P}}{-}\underset{CH_3}{\underset{\mid}{\overset{OH}{\overset{\mid}{C}}}}{-}CH_3 \qquad (6.15)$$

Although many of these addition reactions undoubtedly involve free radicals, some certainly take place by an ionic mechanism and others appear to involve both mechanisms simultaneously. Obviously, further studies are required before these reactions can be fully understood.

Halophosphines will undergo reactions similar to phosphines and phosphites, but the yields of addition products are usually very low owing to side reactions, and so the reactions have little use synthetically. It is often difficult to calculate the energetics of radical reactions involving phosphorus compounds, since bond-energy data are often lacking. However, it has been suggested that, while additions involving initial rupture of phosphorus–hydrogen bonds are probably exothermic, analogous additions involving phosphorus–halogen bonds are much less so; this would explain the low yields and successful competition of side reactions.

6.1.2 *Phosphoranyl radicals*

Phosphonium salts can readily be reduced electrolytically to give phosphines and hydrocarbons, a reaction which almost certainly involves the initial formation of a phosphoranyl radical (6.4). The second step, a reversible decomposition to phosphine and a radical, is well known in a number of other systems.

$$\underset{\text{from cathode}}{R_4P^+ + e} \longrightarrow \underset{(6.4)}{R_4P\cdot} \rightleftharpoons R_3P: + R\cdot \xrightarrow{H\cdot} RH$$

The pyrolysis of certain pentavalent-phosphorus compounds may involve phosphoranyl radicals. For example, heating pentaphenylphosphorane (6.16) in chloroform solution produces tetraphenylphosphonium chloride ($Ph_4P^+Cl^-$) and benzene, probably via an initial decomposition to a phosphoranyl radical.

$$\underset{(6.16)}{Ph_5P} \longrightarrow Ph_4P\cdot + Ph\cdot$$

$$Ph_4P\cdot \xrightarrow{CHCl_3} Ph_4P^+Cl^- \qquad Ph\cdot \xrightarrow{CHCl_3} PhH$$

$$PCl_5 + RH \longrightarrow PCl_3 + RCl + HCl$$

Phosphorus pentachloride will react with hydrocarbons under radical-initiating conditions to give phosphorus trichloride and alkyl chlorides, a reaction which is explained most simply in terms of phosphoranyl radicals. More recently, bis(2,2′-biphenylene)phosphorane (6.17) has been shown to decompose in benzene at room temperature to give (6.19) and (6.20) via the phosphoranyl radical (6.18).

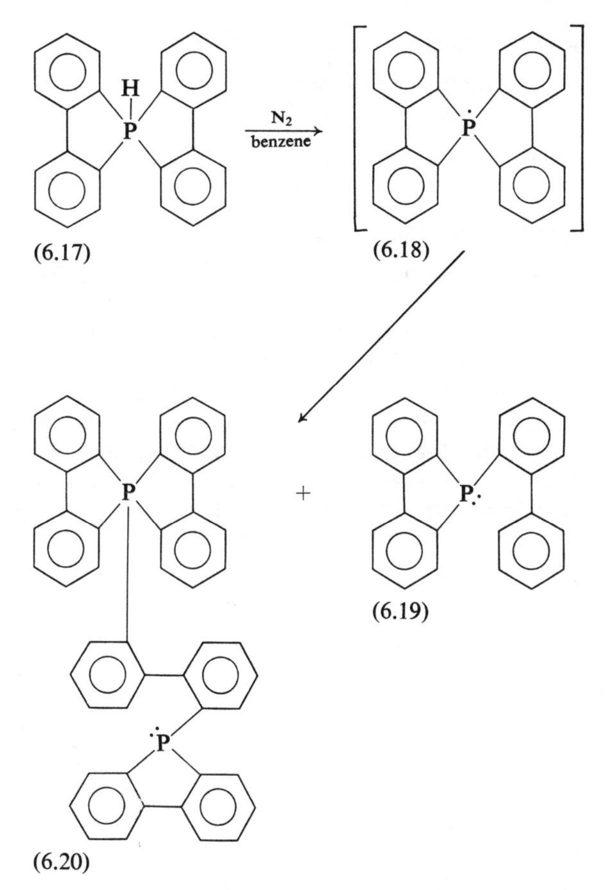

Although there are numerous examples of this type of specialized route to phosphoranyl radicals, by far the commonest method of generating these intermediates is by the reaction of trivalent-phosphorus compounds with free radicals. These reactions will now be studied in some detail. However, it is important to realize that there is often an alternative ionic mechanism for reactions of this type, and a radical reaction should never be assumed without supporting evidence.

(a) *Reaction of trivalent phosphorus with oxygen-containing radicals.* Both phosphines and phosphites react readily with peroxides. Di-tertiary-butyl peroxide and triethyl phosphite (6.21) react at 100°C to give high yields of triethyl phosphate (6.22) and 2,2,3,3-tetramethylbutane (6.23).

$$t\text{-Bu—O—O—}t\text{-Bu} + \underset{(6.21)}{(EtO)_3P:} \longrightarrow \underset{(6.22)}{(EtO)_3P{=}O} + \underset{(6.23)}{(CH_3)_3C\text{—}C(CH_3)_3}$$

$$t\text{-Bu—O—O—}t\text{-Bu} \longrightarrow 2t\text{-BuO}\cdot$$

$$t\text{-BuO}\cdot + (\text{EtO})_3\text{P} \longrightarrow (\text{EtO})_3\dot{\text{P}}\text{—O—}t\text{-Bu} \quad (6.24)$$

$$(\text{EtO})_3\dot{\text{P}}\text{—O—}t\text{-Bu} \longrightarrow (\text{EtO})_3\text{P=O} + t\text{-Bu}\cdot \quad (6.22)$$

$$2t\text{-Bu}\cdot \longrightarrow t\text{-Bu—}t\text{-Bu}$$

The suggested mechanism involves decomposition of the peroxide to alkoxy radicals, followed by attack of these on phosphite to give phosphoranyl radicals (6.24). Decomposition of the phosphoranyl radicals by β-fission leads to triethyl phosphate (6.22) and tertiary-butyl radicals which then dimerize. Recently, evidence has been obtained that the formation of (6.24) is irreversible. Tertiary-butoxy radicals labelled with carbon-14 react with tri(tertiary-butyl) phosphite to give tri(tertiary-butyl) phosphate containing 75 per cent of the original label, rather than the 25 per cent expected in a freely reversible process. Reactions of this type also appear to be stereospecific in that (6.25) and (6.26) react with tertiary-butoxy radicals to give phosphates (6.27 and 6.28) with retention of configuration.

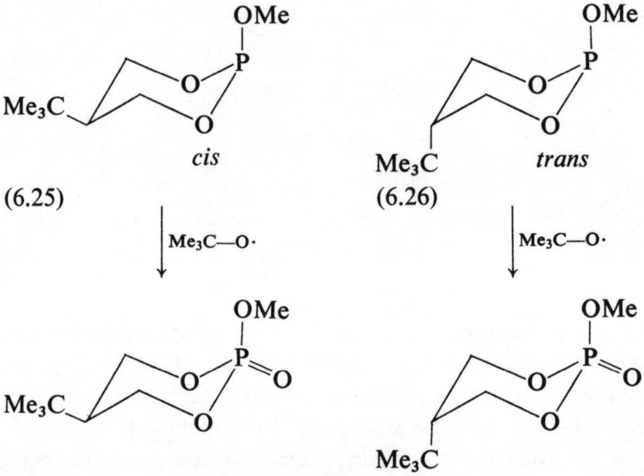

OMe, Me_3C, *cis* (6.25) — OMe, Me_3C, *trans* (6.26)

$\downarrow$ Me_3C—O·

OMe, Me_3C (6.27) — OMe, Me_3C (6.28)

Phosphines probably react in a similar way, but the reaction is often much more complex since, in this case, there are two decomposition pathways available for the initially formed phosphoranyl radical (6.29). It can undergo either β-scission, in a manner analogous to the phosphite example, or α-scission to give a phosphinite (6.30). In forming a P=O bond, there is a gain in enthalpy of 627 kJ mol^{-1} (150 kcal mol^{-1}); in view of this high gain, α-scission should be unfavourable, but a number of examples are known. Tributylphosphine, for instance, reacts with tertiary butyl peroxide to give 80 per cent tertiary-butyl dibutylphosphinite (6.31) (α-scission) and 20 per cent tributylphosphine oxide

$$R_3\dot{P}\text{—}OR' \;(6.29) \xrightarrow{\beta\text{-scission}} R_3P\text{=}O + R'\cdot$$

$$R_3\dot{P}\text{—}OR' \;(6.29) \xrightarrow{\alpha\text{-scission}} \underset{(6.30)}{R_2P\text{—}O\text{—}R'} + R\cdot$$

$$Bu_3P: + t\text{-}BuO\cdot \longrightarrow \underset{\substack{80\% \\ (6.31)}}{Bu_2P\text{—}O\text{—}t\text{-}Bu} + \underset{\substack{20\% \\ (6.32)}}{Bu_3P\text{=}O}$$

(6.32) (β-scission). It has been suggested that these reactions may be bimolecular radical displacements and do not involve an intermediate phosphoranyl radical.

Chemically (or photochemically) initiated reactions of phosphites and phosphines with molecular oxygen are thought to proceed by a similar mechanism. For example, triethylphosphine reacts with oxygen in the presence of a radical initiator to give a complex mixture of all possible oxygenated products.

$$Et_3P + O_2 \longrightarrow Et_3P\text{=}O + Et_2P(\text{=}O)OEt + EtP(\text{=}O)(OEt)_2 + (EtO)_3P\text{=}O$$

$$R\cdot + O_2 \longrightarrow \underset{(6.33)}{R\text{—}O\text{—}O\cdot}$$

$$R\text{—}O\text{—}O\cdot + Et_3P \longrightarrow \underset{(6.34)}{Et_3\dot{P}\text{—}O\text{—}O\text{—}R} \xrightarrow{\beta\text{-scission}} \underset{(6.35)}{Et_3P\text{=}O + RO\cdot}$$

$$R\text{—}O\cdot + Et_3P \longrightarrow Et_3\dot{P}\text{—}O\text{—}R$$

$$Et_3\dot{P}\text{—}O\text{—}R \longrightarrow Et_3P\text{=}O + R\cdot$$

$$Et_3\dot{P}\text{—}O\text{—}R \longrightarrow Et_2P\text{—}OR + Et\cdot$$

$$Et\cdot + O_2 \longrightarrow Et\text{—}O\text{—}O\cdot$$

$$Et_2P\text{—}OR + Et\text{—}O\text{—}O\cdot \longrightarrow Et_2\dot{P}(O\text{—}O\text{—}Et)\text{—}OR$$

$$Et_2\dot{P}(O\text{—}OEt)\text{—}OR \longrightarrow Et_2P(\text{=}O)\text{—}OR + \cdot OEt$$

Scheme **6.9**

The formation of all these products is readily interpreted in terms of α- and β-scission reactions of phosphoranyl radicals. The initiating radical is thought to react with oxygen to give a peroxy radical (6.33), which can then react with the phosphine, via a phosphoranyl radical intermediate (6.34), to give phosphine oxide (6.35) and an alkoxy radical. A combination of α- and β-scission reactions, together with oxygenation of the ethyl radicals produced by α-scission, leads to the products. The principal steps of this mechanism are outlined in Scheme **6.9**.

Trivalent-phosphorus compounds react with nitric oxide to give the phosphorus oxide (6.37) and nitrous oxide, presumably via the phosphoranyl radical (6.36).

$$NO\cdot + (RO)_3P: \longrightarrow \underset{(6.36)}{(RO)_3\dot{P}{-}O^+{=}N^-}$$

$$(RO)_3\dot{P}{-}O^+{=}N^- + NO \longrightarrow \underset{(6.37)}{(RO)_3P{=}O} + N_2O$$

The reaction of quinones with phosphites has been extensively studied, and it appears that this involves an electron transfer from phosphorus to quinone to give phosphinium radical cation (6.38) rather than a phosphoranyl radical. The reaction scheme shown below is considerably simplified.

$$C_6Cl_4O_2 + (RO)_3P \longrightarrow \cdot O{-}C_6Cl_4{-}O^- + \underset{(6.38)}{(RO)_3P\cdot^+}$$

$$\cdot O{-}C_6Cl_4{-}O^- \longrightarrow (RO)_3P^+{-}O{-}C_6Cl_4{-}O^- \longrightarrow (RO)_2P(O){-}C_6Cl_4{-}OH$$

(b) *Reaction of trivalent phosphorus with sulphur-containing radicals.* Thiol radicals react with trivalent-phosphorus compounds in a way analogous to alkoxy radicals. Triethyl phosphite and alkyl thiols, on irradiation with ultraviolet light, give triethyl phosphorothioate (6.39) and the hydrocarbon derived from the thiol.

$$RSH + (EtO)_3P: \longrightarrow \underset{(6.39)}{(EtO)_3P{=}S} + RH$$

This reaction presumably takes place via the thiophosphoranyl radical (6.40) with β-scission of a R—S bond. However, reactions of this type must be interpreted with care since ionic pathways are often available. For example, triethyl

$$RS\cdot + (EtO)_3P: \longrightarrow (EtO)_3\dot{P}{-}S{-}R \quad (6.40)$$

$$(EtO)_3\dot{P}{-}S{-}R \longrightarrow (EtO)_3P{=}S + R\cdot$$

$$PhSH + (EtO)_3P \xrightarrow{\text{radical}} PhH + (EtO)_3P{=}S \quad (6.42) \quad 15\%$$

$$PhSH + (EtO)_3P \xrightarrow{\text{ionic}} PhSEt + (EtO)_2\overset{\overset{O}{\|}}{P}H \quad (6.41) \quad 85\%$$

phosphite and thiophenol undergo simultaneous ionic and radical reactions to give ethylphenyl thioether and diethyl phosphite (**6.41**) (85 per cent) by an ionic mechanism, and benzene and triethyl phosphorothioate (**6.42**) (15 per cent) by a radical mechanism.

The radical contribution to any particular reaction is extremely dependent on the conditions. Disulphides (6.43) normally react with trivalent-phosphorus compounds by an ionic mechanism but, in the presence of a radical initiator, quite a different reaction takes place to give a phosphine sulphide (6.44) and thioether.

$$R'{-}S{-}S{-}R' \longrightarrow 2R'S\cdot \quad (6.43)$$

$$R'S\cdot + R_3P: \longrightarrow R_3\dot{P}{-}S{-}R'$$

$$R_3\dot{P}{-}S{-}R' \longrightarrow R_3P{=}S + R'\cdot \quad (6.44)$$

$$R'\cdot + (R'S)_2 \longrightarrow R'{-}S{-}R' + R'S\cdot$$

A similar situation exists in the reaction between phosphines and elemental sulphur. Under a nitrogen atmosphere, the reaction proceeds smoothly to give a high yield of phosphine sulphide; however, traces of oxygen in the atmosphere cause enormous rate increases, presumably by introducing a radical pathway.

$$8R_3P: + S_8 \longrightarrow 8R_3P{=}S$$

(c) *Reactions of trivalent phosphorus with organic halides.* Ionic reactions between trivalent-phosphorus compounds and alkyl halides were discussed in Chapter 2. However, in the reactions between certain organic halides (particularly poly-halogens) and trivalent phosphorus, there appears to be a contribution from free-radical mechanisms. The subject is still very confused, and the particular pathways are often so finely balanced (even in well-authenticated radical reactions) that a complete change of mechanism can take place with relatively small changes in conditions.

A number of reactions used to prepare tetraarylphosphonium salts may involve free radicals. The photolysis of aryl iodides in the presence of triarylphosphines leads to tetraarylphosphonium salts (6.45), although often in poor yield. Similar products are obtained from the reaction of aryl Grignard compounds with phosphines (6.46) in the presence of oxygen or cobalt salts.

$$Ar_3P: + ArI \xrightarrow{h\nu} Ar_4P^+I^- \quad (6.45)$$

$$Ar_3P: + ArMgX \xrightarrow[\text{or } CoX_2]{O_2} Ar_4P^+ \quad (6.46)$$

Although the reactions of chloroform and bromoform with tertiary phosphorus in the absence of initiators are probably ionic, the situation with carbon tetrachloride is far from clear. Triethyl phosphite reacts with carbon tetrachloride to give the chlorophosphonate (6.47) and ethyl chloride, a reaction which can be interpreted equally well by an ionic or a free-radical mechanism.

$$(EtO)_3P: + CCl_4 \longrightarrow (EtO)_2\overset{O}{\overset{\|}{P}}CCl_3 + EtCl\cdot \quad (6.47)$$

(d) *Reactions involving phosphorus halides.* These are best dealt with separately since, in their radical reactions, they often behave quite differently from other trivalent-phosphorus compounds. In dealing with addition reactions involving phosphinyl radicals (p. 213), it was mentioned that halophosphines usually give poor yields of addition products. Phosphorus trichloride reacts with *p*-xylylene (6.48), even at –78 °C, to give high yields of the polymer (6.49) rather than the simple addition products. This led to the suggestion that, unlike alkyl and aryl phosphines, halophosphines reacted with olefins via phosphoranyl intermediates. It is thought that the initiating radical adds to the *p*-xylylene to form a benzyl radical (6.50), which adds to phosphorus trichloride to form the phosphoranyl

$$CH_2{=}C_6H_4{=}CH_2 + PCl_3 \xrightarrow{-78\,°C} -\!\left(CH_2{-}C_6H_4{-}CH_2{-}\overset{Cl_3}{P}{-}\right)_n$$

(6.48) (6.49)

$$R\cdot + CH_2{=}C_6H_4{=}CH_2 \longrightarrow R{-}CH_2{-}C_6H_4{-}\dot{C}H_2 \quad (6.50)$$

$$R{-}CH_2{-}C_6H_4{-}\dot{C}H_2 + PCl_3 \longrightarrow R{-}CH_2{-}C_6H_4{-}CH_2\dot{P}Cl_3 \quad (6.51)$$

$$R{-}CH_2{-}C_6H_4{-}CH_2\dot{P}Cl_3 + CH_2{=}C_6H_4{=}CH_2 \longrightarrow R{-}CH_2{-}C_6H_4{-}CH_2{-}\overset{Cl_3}{P}{-}CH_2{-}C_6H_4{-}\dot{C}H_2$$

radical (6.51). This radical in turn adds to another molecule of *p*-xylylene, ultimately producing a polymer. The additions of phosphorus trichloride to simple olefins are thought to involve a similar phosphoranyl radical (6.52).

$$R^{1}\cdot + R{-}CH{=}CH_2 \longrightarrow (R)(R^1)CH{-}\dot{C}H_2$$

$$(R)(R^1)CH{-}\dot{C}H_2 + PCl_3 \longrightarrow (R)(R^1)CH{-}CH_2{-}\dot{P}Cl_3$$

(6.52)

The reaction of dihalophosphines (6.53) with dienes has been used extensively in the preparation of phosphorus heterocycles. The reaction here probably proceeds via a concerted cyclic addition rather than a free-radical mechanism.

$$CH_2{=}CH{-}CH{=}CH_2 + \underset{(6.53)}{RPCl_2} \longrightarrow \text{1-R-1,1-dichloro-3-phospholene } (C_4H_6PRCl_2)$$

One interesting and potentially very useful reaction of phosphorus trichloride is that with hydrocarbons, in the presence of oxygen, to give alkyl-substituted halophosphine oxides (6.54). This reaction takes place under very mild conditions and is almost certainly free radical in character. An intensive study of the possible pathways involved has led to the unravelling of what is an extremely complex reaction involving phosphoranyl radicals.

$$2PCl_3 + O_2 + 2RH \longrightarrow 2R\overset{O}{\overset{\|}{P}}Cl_2 + 2HCl$$

(6.54)

6.1.3 *Reactions involving elemental phosphorus*

There has been considerable interest in this subject because reactions involving elemental phosphorus, provided that they can be carried out in reasonable yield and avoid complex separation procedures, are a potential source of cheap phosphorus compounds.

Probably the best-known reaction of phosphorus involves burning white phosphorus in air to give phosphorus pentoxide. This reaction is undoubtedly free radical in nature, as is the alkylation of white phosphorus with olefins in the presence of oxygen.

$$P_4 + 5O_2 \longrightarrow 2P_2O_5$$

Reactions of red (or white) phosphorus with organic halides appear to take place by radical mechanisms, particularly as they require high temperatures and/or the presence of radical initiators and often show an induction period. The yields are variable, and the examples given here are well above the average value.

$$P_4(\text{white}) + EtI(\text{excess}) \xrightarrow[22\,h]{180°C} \underset{49\%}{Et_4P^+I^-}$$

$$P(\text{red}) + 3PhCH_2I \xrightarrow[110°C]{I_2} \underset{81\%}{(PhCH_2)_3P{=}O}$$

$$P(\text{red}) + I(CH_2)_4I \xrightarrow[I_2]{200°C} \underset{40\%}{\text{spiro-}[(CH_2)_4P^+(CH_2)_4]\ I_3^-}$$

$$P_4 + CCl_4 \xrightarrow[I_2]{160°C} Cl_3CPI_2 + PCl_3 + \text{polymers}$$

Elemental phosphorus also reacts with unsaturated compounds, alcohols or phenols, and disulphides (Schemes **6.10–6.12**), probably by radical mechanisms in each case.

$$P(\text{red}) + F_3C{-}C{\equiv}C{-}CF_3 \xrightarrow[I_2]{200°C} \underset{43\%}{(CF_3)_6C_6P_2}$$

Scheme **6.10**

$$P_4 + MeOH \xrightarrow{250°C} (\text{mixtures of phosphines})$$

Scheme **6.11**

$$P_4 + 6RSSR \xrightarrow{200°C} \underset{70\%}{4(RS)_3P:}$$

Scheme **6.12**

The other typical reaction of elemental phosphorus is that with nucleophilic reagents. The reaction of white phosphorus with aqueous sodium hydroxide to give phosphine (PH_3) and phosphorus oxyacid salts is probably the best known of these. White phosphorus reacts readily with metal alkyls to give deep red solutions of what are generally thought to be organophosphides. These organophosphides will react with electrophiles to give organophosphorus compounds in yields varying from traces to 80 per cent. The reason for the variable, and

$$RLi + P_4 \longrightarrow [\text{organophosphide}]$$

$\xrightarrow{R'X}$ $RPR'_2 + R_2PR'$ $\sim 40\%$

$\xrightarrow{H_2O}$ RPH_2 + polymers 0—36%

often low, yields is the multitude of pathways available from the reaction of a nucleophile with a P_4 pyramid (6.55). This pyramid has six phosphorus–phosphorus bonds to break before it can yield any products containing only one phosphorus atom.

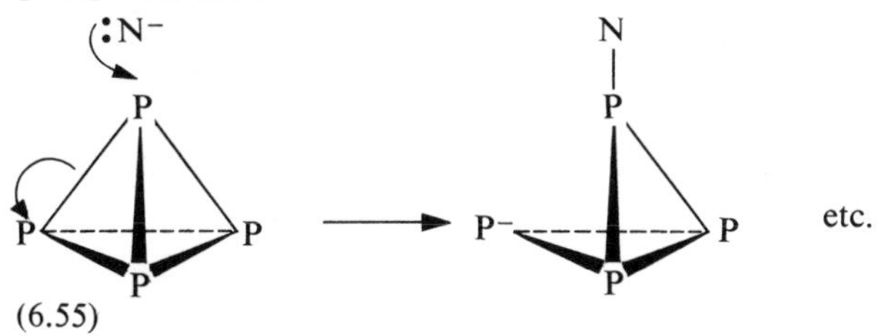

(6.55)

6.2 Compounds containing a phosphorus–phosphorus bond

6.2.1 *Diphosphines*

The diphosphines discussed here are those containing a phosphorus–phosphorus bond (6.56). When two phosphorus atoms are separated by one or more carbon atoms, the chemistry of the resulting *diphosphines* closely resembles that of the trivalent-phosphorus compounds discussed in Chapter 2. However, the introduction of a phosphorus–phosphorus link modifies their chemistry to such an extent that these compounds are best studied separately.

$$R_2P\text{—}PR_2$$

(6.56)

$$>\ddot{P}\text{—}\ddot{P}< \;\longleftrightarrow\; >P^{+}\text{=}\bar{P}< \;\longleftrightarrow\; >P^{-}\text{=}\overset{+}{P}< \;\longleftrightarrow\; >P\equiv P<$$

(a) (b) (c) (d)

(6.57)

Since each phosphorus atom has both a lone pair of electrons and available d-orbitals which are thought to participate in π-bonding, diphosphines are probably best represented by a series of canonical forms (6.57). Some evidence for the overlap of the lone pair and the d-orbital is available from ultraviolet spectra, and nuclear magnetic resonance can be very usefully employed in determining the magnitude of the rotational barrier. Since *cis* and *trans* forms of the diphosphines have not been separated, the barrier must be low compared to that in olefins (approximately 252 kJ mol^{-1} (60 kcal mol^{-1})), which are presumably equivalent to a 100 per cent contribution of the forms (6.57b and c).

$$(CH_3)_2P{-}N(CH_3)_2 + (CH_3)_2PH \longrightarrow (CH_3)_2P{-}P(CH_3)_2 + HN(CH_3)_2$$

(6.58) (6.59)

The reaction of the aminophosphine (6.58) with dimethylphosphine to give tetramethyldiphosphine (6.59) and dimethylamine, has been quoted as evidence for contributions from the canonical form (6.57d). *N,N*-dimethyl dimethylphosphinous amide (6.58) can only contain contributions from doubly bonded canonical forms analogous to (6.57b and c), since d-orbitals are not available on a nitrogen atom to take part in bonding analogous to that in (6.57d). However, diphosphines show chemical behaviour consistent with a dominant contribution from the form (6.57a), with a phosphorus–phosphorus single bond.

Attempts have been made to determine the stereochemistry about the phosphorus–phosphorus bond of diphosphines. The *trans* form (6.60) might be expected to be most stable, but although this appears to be true for tetraiododiphosphine ($PI_2{-}PI_2$) in the solid phase, gas-phase vibrational spectral studies on diphosphine itself suggest a gauche arrangement (6.61).

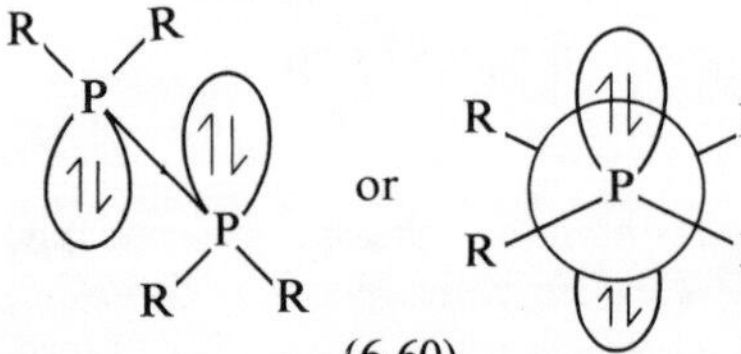

R R P P R R or R R P R R

(6.60)

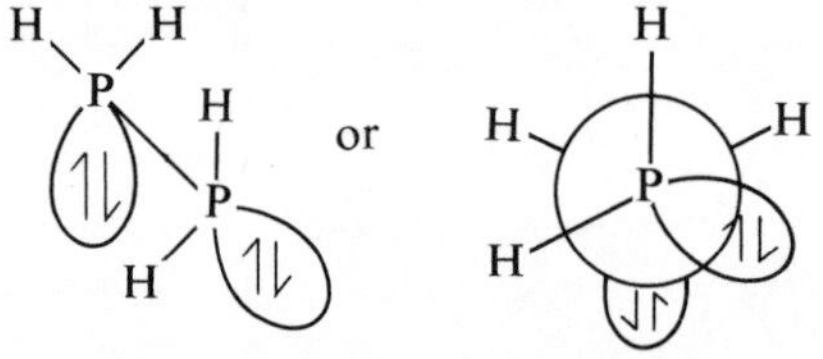

H H P H P H or H H P H H

(6.61)

$$\begin{matrix} R \\ R^1 \end{matrix}{>}P{-}P{<}\begin{matrix} R \\ R^1 \end{matrix}$$

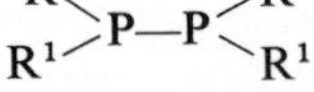

(6.62)

Since, unlike nitrogen compounds, phosphorus(III) compounds do not undergo pyramidal inversion at temperatures below about 100 °C (see Chapter 1, p. 26), resolution of asymmetric diphosphines (6.62) should be possible. This has been achieved for both diphosphines and diphosphine disulphides.

(a) *Formation.* Methods of synthesizing phosphorus–phosphorus bonds are fairly restricted, and the main ones are summarized in Table 36 (p. 228). As might be expected, the methods are mainly analogous to those used in the preparations of cyclopolyphosphines (see p. 235). The reaction of Grignard reagents with Cl_3PS (method 7) might be expected to give tertiary phosphine sulphides (R_3PS) rather than diphosphine disulphides. The mechanism of formation of the latter compounds is still obscure, although recently phosphinidene sulphide (RP=S) has been suggested as an intermediate.

Method 10 has been shown to involve a metal–halogen exchange as the first step, followed by a nucleophilic substitution at the phosphorus atom to give the diphosphine (6.63). The ethylene is probably formed by elimination from the lithium alkyl (6.64).

$$R_2PLi + Br{-}CH_2{-}CH{-}Br \longrightarrow R_2PBr + LiCH_2CH_2Br$$

$$R_2PBr + R_2PLi \longrightarrow \underset{(6.63)}{R_2P{-}PR_2} + LiBr$$

$$\underset{(6.64)}{LiCH_2{-}CH_2{-}Br} \longrightarrow CH_2{=}CH_2 + LiBr$$

Methods of synthesis starting from compounds already containing phosphorus–phosphorus bonds are rare, since the common reaction of such compounds is rupture of this bond.

(b) *Reactions.* As might be expected, diphosphines undergo many of the reactions characteristic of trivalent-phosphorus compounds (see Chapter 2); however, the dominant factor in their reactivity is the relative weakness of the phosphorus–phosphorus bond. This has a bond strength of about 209 kJ mol^{-1} (50 kcal mol^{-1}), and so will be fairly easily broken. Yet many organic diphosphines are thermally stable at temperatures as high as 300 °C. Usually, it is not until the attack of a reagent at only one of the phosphorus atoms makes the phosphorus–phosphorus bond unsymmetrical that bond fracture takes place easily. On heating to temperatures greater than 300 °C, diphosphines undergo disproportionation to give tertiary phosphines (6.65), cyclopolyphosphines (6.66), polymers and elemental phosphorus.

$$R_2P{-}PR_2 \xrightarrow{>300^\circ} \underset{(6.65)}{R_3P} + \underset{(6.66)}{(RP)_n} + \text{polymers} + P \qquad \underset{(6.67)}{Ph_2P{-}PPh_2}$$

Although tetraphenyl diphosphine (6.67) may be distilled at 260 °C at 1 mm pressure without decomposition, it does appear to dissociate into free radicals to some extent even at temperatures as low as 180 °C. Since diphosphines will also dissociate readily on irradiation with ultraviolet light, they have been frequently used as sources of phosphinyl radicals ($R_2\ddot{P}\cdot$) (see p. 213). Particular care must be taken when interpreting diphosphine reactions, since a free-radical mechanism is often a possibility.

Table 36 Preparation of diphosphines

Eliminations

1. $2R_2PX + \text{metal (Hg, Na, Li, etc.)} \xrightarrow[\text{heat}]{\text{dioxan}} R_2P\text{—}PR_2 + \text{metal halide}$
 (R = Et; b.p. 221 °C at 760 mm; 70%)

2. $R_2PX + R_2PH \xrightarrow[\text{hexane, reflux, 4 h}]{-HX} R_2P\text{—}PR_2$
 (R = Ph; m.p. 120 °C) good yield

3. $R_2PX + R_2PM \; (M = Li, Na, K) \xrightarrow{\text{THF}} R_2P\text{—}PR_2 + MX$
 (good yield)

From elemental phosphorus

4. $P \text{ (red or white)} + I_2 \longrightarrow I_2P\text{—}PI_2$
 (poor yield)

5. $P_4 \text{ (white)} + Ph_4Sn \xrightarrow{\text{sealed tube}} (Ph_3Sn)_2P\text{—}P(SnPh_3)_2$

Miscellaneous

6. $(PhP)_4 + 4Na \longrightarrow 2Ph\text{—}P^-\text{—}P^-\text{—}Ph \; Na^+Na^+ \xrightarrow{2RX} (R)(Ph)P\text{—}P(R)(Ph) + 2NaX$
 (good method for making unsymmetrically substituted diphosphines)

7. $RMgBr + Cl_3P{=}S \longrightarrow R_2P(=S)\text{—}P(=S)R_2 \xrightarrow[\text{or } R_3P:]{\text{metal}} R_2P\text{—}PR_2$
 (good method: but does not work if R is aryl or sterically hindered alkyl)

8. $Me_2PH + Me_2N\text{—}PMe_2 \longrightarrow Me_2P\text{—}PMe_2 + Me_2NH$
 (85%)

9. $Ca_3P_2 + H_2O \longrightarrow PH_2\text{—}PH_2$ (poor yield)

10. $R_2PLi + Br\text{—}CH_2\text{—}CH_2Br \xrightarrow[\text{heat}]{\text{toluene}} R_2P\text{—}PR_2 + CH_2{=}CH_2 + 2LiBr$
 (R = Et; 81%)

Diphosphines have been compared to halogens, and this *pseudohalogen* concept (first suggested by Anton Burg) has been used to some effect in predicting reactivity. For example, diphosphines will add to olefins to give diphosphines in which the phosphorus atoms are separated by two carbon atoms (6.68).

$$R_2P{-}PR_2 + CH_2{=}CH_2 \longrightarrow \underset{(6.68)}{R_2P{-}CH_2{-}CH_2{-}PR_2}$$

Diphosphines are readily oxidized in air, to the extent that tetraphenyldiphosphine spread on a filter paper in the open laboratory will get quite hot. The product from this oxidation is tetraphenyldiphosphine dioxide (6.69) a compound which, drawn in its dipolar form (6.70), contradicts Pauling's adjacent-charge rule. Some confusion exists in the literature about the diphosphine oxides, because a monoxide (6.71) also exists and this has a similar melting point to the dioxide (6.69). Early papers in particular should be treated with extreme caution since, in a number of cases, incorrect assignments have been made.

$$Ph_2P{-}PPh_2 \xrightarrow{\text{air}} \underset{(6.69)}{Ph_2\overset{O}{\overset{\|}{P}}{-}\overset{O}{\overset{\|}{P}}Ph_2} \qquad \underset{(6.70)}{Ph_2\overset{O^-}{\overset{|}{P^+}}{-}\overset{O^-}{\overset{|}{P^+}}Ph_2} \qquad \underset{(6.71)}{Ph_2\overset{O}{\overset{\|}{P}}{-}PPh_2}$$

Neutral hydrolysis is very slow at room temperature but, at higher temperatures, or under alkaline conditions, the phosphorus–phosphorus bond is rapidly broken. Tetrakis(trifluoromethyl)diphosphine (6.72) is hydrolysed to give three moles of fluoroform (6.73) from each mole of diphosphine under alkaline conditions, whereas each mole of diphosphine gives only two moles of fluoroform under acid hydrolysis. It is thought that bis(trifluoromethyl)-phosphine (6.74) and its phosphine oxide (6.75) are formed initially under either acid or alkaline conditions. Bis(trifluoromethyl)phosphine oxide (6.75) is known to be hydrolysed by either acid or alkali to give two moles of fluoroform and one mole of phosphorous acid (6.76). However, while bis(trifluoromethyl)-phosphine gives one mole of fluoroform under alkaline hydrolysis conditions, it gives trifluoromethylphosphine (6.77) and no fluoroform under acid conditions.

$$\underset{(6.72)}{(CF_3)_2P{-}P(CF_3)_2} \xrightarrow{OH^-} \underset{(6.73)}{3CHF_3}$$

$$\underset{(6.72)}{(CF_3)_2P{-}P(CF_3)_2} \xrightarrow{H^+} \underset{(6.73)}{2CHF_3}$$

The phosphorus–phosphorus bond in diphosphines is also cleaved by halogens, alkali metals, metal alkyls and hydrogenation catalysed by Raney nickel, although not by lithium tetrahydroaluminate. Unfortunately, lithium tetrahydroaluminate does cleave the phosphorus–phosphorus bond in diphosphine dioxides, and so this cannot be used as a method of preparation for diphosphines.

$$(CF_3)_2P{-}P(CF_3)_2 \xrightarrow[\text{or } OH^-]{H^+} \underset{(6.75)}{CF_3\overset{\overset{\large O}{\|}}{P}{-}H} + \underset{(6.74)}{(CF_3)_2PH}$$

(6.75) $\xrightarrow{H^+ \text{ or } OH^-}$ $\underset{(6.76)}{2CHF_3 + H_3PO_3}$

(6.74) $\xrightarrow{H^+}$ $\underset{(6.77)}{CF_3PH_2}$ + F^- + CO_3^{2-}

(6.74) $\xrightarrow{OH^-}$ CHF_3 + F^- + CO_3^{2-} + acids of phosphorus

$$R_2PH \xleftarrow[H_2]{Ni} R_2P{-}PR_2 \xrightarrow{X_2} R_2PX$$

$R_2P{-}PR_2 \xrightarrow{R'Li} R_2PLi + R_2PR'$

$R_2P{-}PR_2 \xrightarrow{\text{M (alkali metal)}} R_2P^-M^+$

Diphosphines are weakly basic, but reaction with strong acids usually leads to the breaking of the phosphorus–phosphorus bond. For example, with hydrochloric acid, a dialkylphosphonium salt (6.78) and a chlorophosphine (6.79) are formed. An exception here is tetrakis(trifluoromethyl)diphosphine which apparently reacts with hydrochloric acid without fracture of the diphosphine link.

$$R_2P{-}PR_2 + 2HCl \longrightarrow \underset{(6.78)}{R_2P^+H_2Cl^-} + \underset{(6.79)}{R_2PCl}$$

Reactions with alkyl halides are similar, but retention of the diphosphine link is more common. Although tetraphenyldiphosphine (6.80) and tetracyclohexyldiphosphine (6.81) undergo cleavage to give a mixture of phosphines, most

$$\underset{(6.80)}{Ph_2P{-}PPh_2} + RX \longrightarrow Ph_2P{-}R + Ph_2PX$$

$$\underset{(6.81)}{(C_6H_{11})_2P{-}P(C_6H_{11})_2} + RX \longrightarrow (C_6H_{11})_2PR + (C_6H_{11})_2PX$$

$$R_2P{-}PR_2 + R'X \longrightarrow \underset{(6.82)}{\underset{\underset{\large R'}{|}}{R_2P^+}{-}PR_2\ X^-}$$

tetraalkyldiphosphines form a monophosphonium salt (6.82) with retention of the phosphorus–phosphorus bond.

Diphosphines can act as electron donors in a variety of reactions. They will react with Lewis acids to form either mono- (6.83) or di-adducts (6.84), and a diadduct (6.85) with diborane has been used to stabilize diphosphine, which is spontaneously inflammable in the free state.

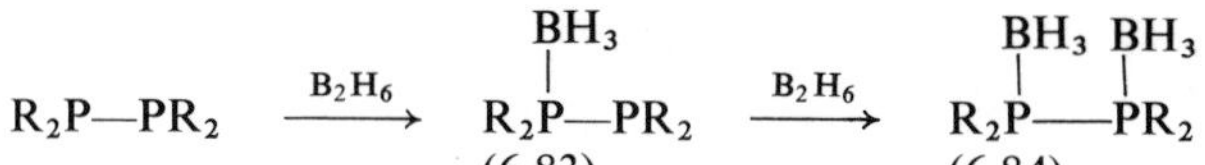

$$R_2P\text{—}PR_2 \xrightarrow{B_2H_6} \underset{(6.83)}{R_2\overset{BH_3}{P}\text{—}PR_2} \xrightarrow{B_2H_6} \underset{(6.84)}{R_2\overset{BH_3}{P}\text{—}\overset{BH_3}{P}R_2}$$

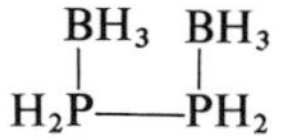

$$\underset{(6.85)}{H_2\overset{BH_3}{P}\text{—}\overset{BH_3}{P}H_2}$$

Diphosphines display at least three different types of ligand behaviour towards transition metals. The diphosphine can act as a bidentate ligand to two separate metal atoms, for example (6.86) and (6.87). Alternatively, both electron pairs can be donated to one metal atom with the formation of a three-membered ring (6.88), although examples of this are rare. Finally, the diphosphine link appears to be broken in some cases, and complexes (6.89) are formed which have phosphino-bridges. There is some evidence that these compounds also have metal–metal bonds.

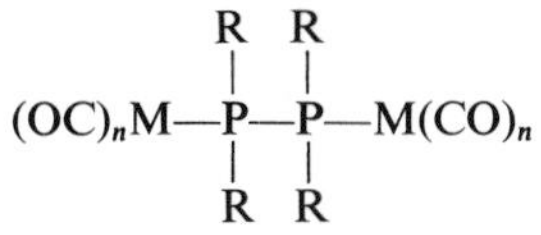

$$(OC)_nM\text{—}\underset{R}{\overset{R}{P}}\text{—}\underset{R}{\overset{R}{P}}\text{—}M(CO)_n \quad (6.86)$$

$$(ON)_2Fe(\leftarrow PR_2\text{—}PR_2\rightarrow)_2Fe(NO)_2 \quad (6.87)$$

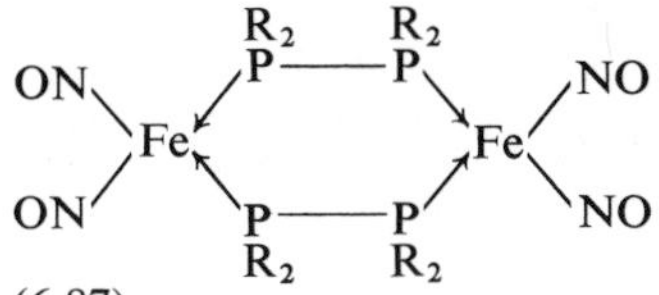

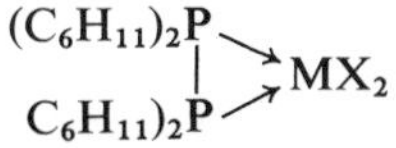

$$(C_6H_{11})_2P\text{—}P(C_6H_{11})_2 \rightarrow MX_2 \quad (6.88)$$

$$(OC)_nM(\mu\text{-}PR_2)_2M(CO)_n \quad (6.89)$$

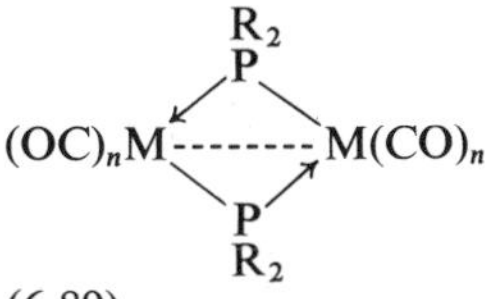

6.2.2 *Higher polyphosphines*

Little attention has been given to the chemistry of compounds containing a chain of more than two connected phosphorus atoms. One problem is that the methods of synthesis available are based on analogy with those used for diphosphines and, in many cases, they are unsatisfactory for polyphosphines. In addition to this, polyphosphines appear readily to undergo disproportionation, often at quite low temperatures.

The methods used for diphosphine synthesis (Table 36) have been applied to the polyphosphines, with varying success. Two moles of diphenylphosphine react with one mole of bromodiphenylphosphine in a vacuum at –10 °C to give the triphosphine disalt (6.90). The free phosphine can be obtained by treatment of (6.90) with triethylamine; it is rather unstable thermally, decomposing at 70 °C.

$$2Ph_2PH + PhPBr_2 \xrightarrow[\text{vacuum}]{-10\,^{\circ}C} Ph_2P\langle\begin{smallmatrix}\overset{H}{P^+}Ph_2Br^- \\ \underset{H}{P^+}Ph_2Br^-\end{smallmatrix} \quad (6.90)$$

$$\xrightarrow{2Et_3N} Ph_2P\langle\begin{smallmatrix}PPh_2 \\ PPh_2\end{smallmatrix}$$

The triphenyltriphosphine (6.91) has been prepared by treatment of dibromophenylphosphine with lithium hydride, presumably via reduction to phenylphosphine.

$$3PhPBr_2 + 6LiH \xrightarrow[\text{benzene}]{5\,^{\circ}C} Ph{-}P\langle\begin{smallmatrix}\overset{H}{P}Ph \\ \underset{H}{P}Ph\end{smallmatrix} \quad (6.91)$$

$$PBr_3 + 3NaPPh_2 \xrightarrow{\text{benzene}} \text{polymeric material} \qquad P(PPh_2)_3 \quad (6.92)$$

Attempts to make branched-chain polyphosphines have failed because these compounds are even more susceptible to disproportionation than their straight-chain analogues. The reaction of phosphorus tribromide with sodium diphenylphosphide leads to mixtures of polymers, rather than the expected tetraphosphine (6.92), although this may be formed initially. From the few studies that have been carried out, it appears that branched-chain isomers are much more liable to disproportionate than the equivalent straight-chain analogues; for example, the tetraphosphine (6.93) has been prepared.

$$CF_3\overset{H}{P}{-}\overset{CF_3}{P}{-}\overset{CF_3}{P}{-}\overset{H}{P}CF_3 \quad (6.93)$$

As might be expected, the polyphosphines are strong reducing agents. They are often spontaneously inflammable, and the triphosphine (6.91) explodes when treated with concentrated nitric acid.

6.2.3 *Cyclopolyphosphines*

Like all elements in the second row of the periodic table, phosphorus shows very poor p_π–p_π overlap (see Chapter 1, p. 19). This leads to catenation rather than p_π-multiple-bond formation, and compounds like R—P=O, R—P=CR_2, R—P=N—R and R—P=S are either unknown, or rapidly dimerize. A large number of structures of this type can be found in the literature of as recently as ten years ago, but it seems likely that, on reinvestigation, these will turn out to be dimers, or higher polymers. (However see p. 20.)

R—P=P—R
(6.94)

The cyclopolyphosphines are closely related to compounds of this type since, until 1959, they were thought to have structures (6.94) involving p_π bonding between phosphorus atoms. More recently, they have been shown to contain rings of three, four, five and six phosphorus atoms.

(a) *Formation.* Cyclopolyphosphines (6.95) can be considered to be cyclic polymers of the phosphorus analogues (6.96) of carbenes. Indeed, any attempt to prepare the carbene analogues (6.96) leads to cyclopolyphosphines unless a suitable trap is present. Because of this, it is possible that most preparations of cyclopolyphosphines involve the intermediacy of ($R\ddot{P}$:).

The range of cyclopolyphosphines so far prepared is fairly restricted, and the majority are listed in Table 37. The methods available for their preparation are summarized in Table 38. In many cases, the yields are low; but the starting materials are often readily available, and so preparation does not usually present any problems.

$(RP)_n$ (6.95) $R—\ddot{P}$: (6.96) $(RP)_n \xrightarrow{160\,°C} R\ddot{P}$: (6.96)

The carbene analogue (6.96) has been trapped (see p. 234) from a number of these reactions; for example, from the reactions of tributylphosphine with iodine. However, probably the most convenient way of generating (6.96) is by the pyrolysis of the corresponding cyclopolyphosphine.

(b) *Reactions.* Since cyclopolyphosphines and their monomers ($R\ddot{P}$:) appear to be interconvertible, it is often difficult to tell which species is taking part in any particular reaction. However, phosphorus analogues of carbenes do appear to be disappointingly unreactive and, except in the presence of some particularly inviting reagent, rapidly telomerize to the cyclopolyphosphines. To some extent, this impression of stability (compared with say, nitrenes) may be overemphasized. Both nitrenes and their phosphorus analogues, in comparison with carbenes, have an extra pair of unbound electrons and, since, in the majority of their reactions, carbenes act as electrophiles, this type of reactivity would be reduced. In addition to this, phosphorus is much less electronegative than nitrogen and so should be even more nucleophilic. That this is the reason for the apparent stability of univalent phosphorus could be verified by careful choice of reagents and phosphorus substituents.

Table 37 Compounds containing *n*-membered phosphorus rings

n	*Compound*
3	P_4 vapour
	$K_2(PhP)_3$
	$(PC_2F_5)_3$
4	$(PEt)_4$
	$(n\text{-PrP})_4$
	$(iso\text{-PrP})_4$
	$(n\text{-BuP})_4$
	$(t\text{-BuP})_4$
	$(n\text{-}C_8H_{17}P)_4$
	$(PhP)_4$
	$(CF_3P)_4$
5	$(MeP)_5$
	$(CF_3P)_5$
	$(PhP)_5$
6	$(PhP)_6$
	$(p\text{-}ClC_6H_4P)_6$

Some evidence exists for the existence of rings with $n > 6$

A number of reactions are known which appear to be best explained as involving univalent phosphorus. The pyrolysis of cyclopolyphosphines in the presence of diethyldisulphide leads to the formation of (6.97). An identical reaction takes place with univalent phosphorus generated by other methods. Phenylphosphorus (phenylphosphinidene) will also react with benzil to give the phosphorane (6.98).

$$(PhP)_n \xrightarrow{160\,°C} [Ph\ddot{P}:] \xrightarrow{Et_2S_2} PhP(SEt)_2 \quad (6.97)$$

$$[Ph\ddot{P}:] + PhC(=O)C(=O)Ph \longrightarrow$$

Ph Ph O O P O O Ph Ph Ph

(6.98)

Table 38 Preparation of cyclopolyphosphines

Dehalogenation

1. $n\text{RPCl}_2 + n\text{Mg} \longrightarrow (\text{RP})_n + n\text{MgCl}_2$

2. $n\text{RPX}_2 + n\text{Hg} \longrightarrow (\text{RP})_n + n\text{HgX}_2$
   X = I or Br

3. $n\text{RPCl}_2 + n\text{RPH}_2 \xrightarrow[\text{toluene}]{\text{reflux}} 2(\text{RP})_n + n\text{HCl}$

4. $n\text{RPCl}_2 + \text{LiAlH}_4 \xrightarrow[\text{ether}]{\text{reflux}} (\text{RP})_n$

5. $n\text{RPF}_2 \xrightarrow[\text{sealed tube}]{20\text{—}40\,^\circ\text{C}} (\text{RP})_n$

6. $n\text{RPH}_2 + n\text{R}_3\text{PCl}_2 \xrightarrow[\text{CCl}_4]{0\,^\circ\text{C}} 2(\text{RP})_n$

Other methods

7. $n\text{RP(NR}_2)_2 + n\text{RPH}_2 \longrightarrow 2(\text{RP})_n$

8. $4\text{PhPH}_2 + 4\text{R}_2\text{Hg} \longrightarrow (\text{PhP})_4 + 4\text{Hg} + 8\text{RH}$

9. $n\text{R}\overset{\text{O}}{\overset{\|}{\text{P}}}\text{H}_2 \xrightarrow[\text{vacuum}]{\text{heat}} (\text{RP})_n$ (low yield)

10. $n(\text{CF}_3)_2\text{P—P(CF}_3)_2 \xrightarrow{350\,^\circ\text{C}} (\text{CF}_3\text{P})_n + n(\text{CF}_3)_3\text{P}$

11. $n\text{Bu}_3\text{P} \xrightarrow[\text{I}_2]{250\,^\circ\text{C}} (\text{BuP})_n$ (very low yield)

12. $\text{R—MgBr} + \text{P}_4$ (white phosphorus) $\longrightarrow (\text{RP})_4$

Reactions with carbon–carbon multiple bonds are not common, but butylphosphorus (6.99) will react with diphenylacetylene to give *P*-butyltetraphenylphosphole (6.100). Cyclopolyphosphines will react with substituted butadienes, but compounds (6.101) containing phosphorus–phosphorus links are found amongst the products; this suggests that intermediates other than simple univalent phosphorus may be involved.

$$BuP: + 2\, PhC{\equiv}CPh \longrightarrow \text{1-butyl-2,3,4,5-tetraphenylphosphole}$$

(6.99) (6.100)

$$(RP)_n + R'C(=CH_2)C(=CH_2)R' \xrightarrow{180^\circ C} \text{(1,2-diphosphacyclohexene, P–P, R on each P)} + \text{(phospholene, P–R)}$$

(6.101)

The stability of cyclopolyphosphines in air depends on the substituents, but generally they are oxidized to give polymeric materials of composition $(RPO_2)_n$ ($(CF_3P)_4$ is spontaneously inflammable). Similarly, conditions required for hydrolysis vary considerably from compound to compound, and the products obtained are extremely dependent on the conditions of hydrolysis. Hydrolysis of the trifluoromethyl tetramer (6.102) has been extensively studied, and some of the products obtained under various conditions are shown below.

$$CF_3P(H){-}P(CF_3){-}P(H)CF_3 + CF_3P(H){-}P(CF_3){-}P(CF_3){-}P(H)CF_3 \xleftarrow{\text{alcohols}} (CF_3P)_4 \text{ (6.102)} \rightarrow CF_3PH_2 + CF_3P(=O)(H)OH \text{ ; } CF_3P(H){-}P(H)CF_3$$

The phosphorus–phosphorus bond of cyclopolyphosphines is readily broken by alkali metals to give metal phosphides (6.103), or by halogens to give dihalophosphines (6.104). Reactions occurring with retention of the phosphorus ring are rare. Although ethylphosphorus tetramer (6.105) will react with one mole of methyl iodide to give the phosphonium salt (6.106), most other

$$(RP)_n + 2nNa \longrightarrow nRP^{2-}2nNa^+ \quad (6.103)$$

$$(RP)_n + nX_2 \longrightarrow nRPX_2 \quad (6.104)$$

$$(EtP)_4 \text{ (6.105)} + MeI \longrightarrow [(EtP)_4CH_3]^+I^- \quad (6.106)$$

$$(PhP)_n + Ni(CO)_4 \longrightarrow (PhP)_4Ni(CO)_3$$
(6.107)

$$(PhP)_n + Mo(CO)_6 \longrightarrow (PhP)_5Mo(CO)_5$$
(6.108)

cyclopolyphosphines undergo complete breakdown of the ring under similar conditions. However, almost all polyphosphines will react with metal carbonyls with ring retention, although, depending on the metal involved, the ring size may change (e.g. 6.107 and 6.108).

(c) *Structures*. Phosphobenzene $(PhP)_n$ was originally formulated as a dimer (6.109) in 1877 by Michaelis. This structure was accepted generally for eighty years until molecular-weight determination in solution suggested it was a tetramer $(PhP)_4$. However, this apparently simple explanation is complicated by phosphobenzene being obtainable in at least four distinct forms. This was originally thought to be due to polymorphism, but difficulties of interconversion, even by preferential seeding, did not support this explanation.

Ph—P=P—Ph
(6.109)

Two of the four forms are probably polymeric mixtures, since they are extremely insoluble, but the other two forms (generally called *A* m.p. 148–50 °C, and *B* m.p. 195–9 °C) do appear to be distinct molecules. Both *A* and *B* appear to give solutions which contain $(PhP)_4$, although solutions of *A* always give *A* back and, similarly, solutions of *B* always give *B* back, even when seeded by crystals of the other isomer.

The problem has been partially resolved by X-ray crystallographic studies. In the solid form, *A* has been shown to contain a five-membered ring (6.110), while form *B* has a six-membered ring (6.111). These studies have also shown that *B* itself can exist in several different crystalline modifications; the one which has been studied has a chair conformation. Much more work is required before a clear picture of the structures of these compounds in solution can be obtained.

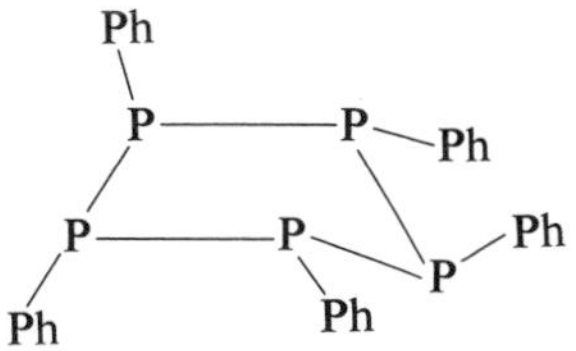

Ph
P — P—Ph
P — P — P—Ph
Ph Ph
(6.110)

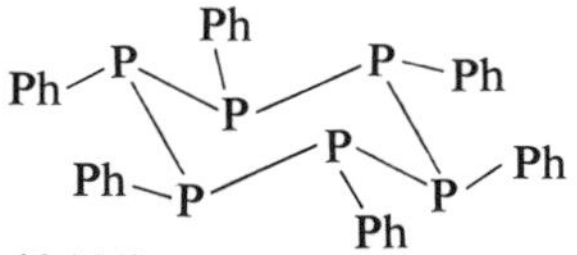

Ph
Ph—P P P—Ph
Ph—P P P—Ph
Ph
(6.111)

Fortunately, other cyclopolyphosphines do not appear to be as complex, although this may in part be due to the fact that they have been studied less intensively. Trifluoromethylphosphorus can be obtained in two distinct forms, X-ray studies have shown these to be a tetramer (6.112) and pentamer (6.113). The tetramer is non-planar with staggered trifluoromethyl groups, presumably

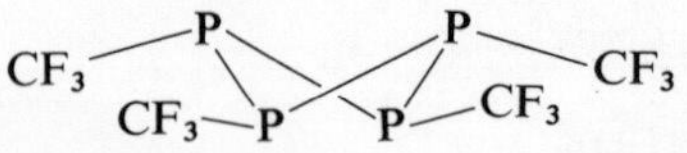

(6.112) 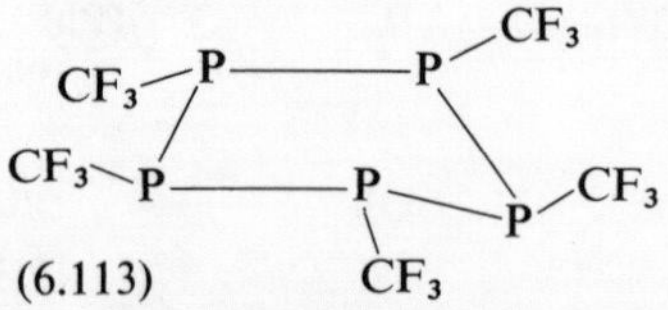 (6.113)

because this configuration reduces torsional strain. The pentamer is more nearly planar, with eclipsed trifluoromethyl groups at the phosphorus atoms 3 and 4.

X-ray structures of alkyl cyclophosphines are not available, but mass-spectroscopic studies suggest that what were originally thought to be tetramers (e.g. $(EtP)_4$) may also contain pentameric structures.

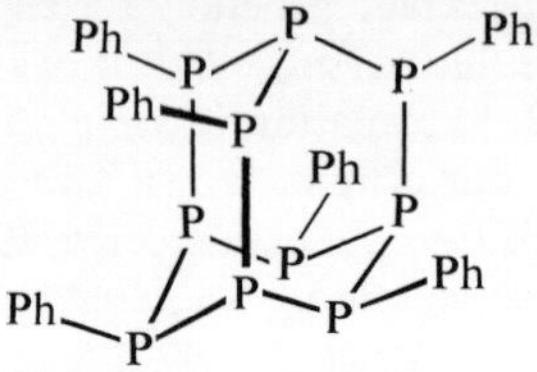

(6.114)

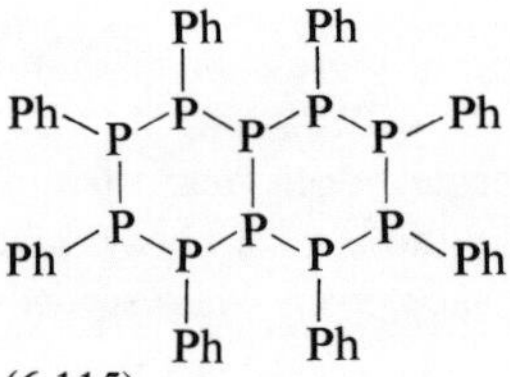

(6.115)

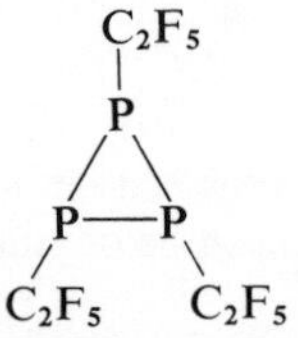

(6.116) 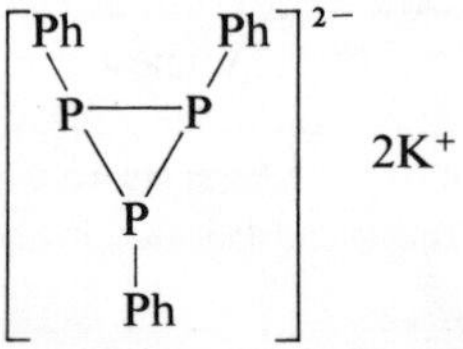 (6.117)

More complex cyclic polyphosphine structures are also known (e.g. 6.114 and 6.115), but one of the most interesting recent developments has been the isolation of compounds (6.116) and (6.117) containing three-membered phosphorus rings. Compound (6.116) was prepared, together with $(C_2F_5P)_4$, by the reaction of $C_2F_5PI_2$ and mercury, and the three-membered cyclic structure was suggested on the basis of mass spectra and cryoscopic molecular-weight determinations. The dianion (6.117) is formed by the reaction of potassium metal with the phenylphosphorus pentamer. The comparative stability of (6.117) has been related to the possibility of a delocalized orbital, composed of d-orbitals and containing two electrons, giving the compound Hückel aromatic character.

Problems

6.1 Discuss the structure of phosphoranyl radicals ($R_4P\cdot$) and show how your proposed structure can be reconciled with the fact that optically active phosphines react with tertiary butoxy radicals to give phosphine oxides with retention of configuration.
[Bentrude and Wielesek (1969); Bentrude, Hargis and Rusek (1969).]

6.2 Carbene analogues of a large number of elements have been prepared. Describe the expected electronic structures of the nitrogen, phosphorus, oxygen and sulphur analogues of carbenes and, in terms of these structures, predict their comparative chemistries.
[Nefedov and Manakov (1966), p. 1033. See also Abramovitch and Davis (1964).]

6.3 Suggest a reasonable mechanism for each of the following reactions.

(a) $C_6H_{11}PH^-K^+$ + $BrCH_2CH_2Br$ ⟶ $CH_2{=}CH_2$ + (cyclohexyl-P)$_4$ (tetracyclohexylcyclotetraphosphane)

(b) (EtO)$_3$P dioxaphospholane ring bearing Ph, NC / Ph, CN substituents $\xrightarrow[h\nu]{RCH=CHR}$ cyclopropane (C bearing Ph and CN, RHC—CHR) + PhCOCN + $Et_3P{=}O$

[Petrellis and Griffin (1968).]

(c) $(MeO)_2POCH_2COR \xrightarrow{h\nu} (MeO)_2P(=O)CR{=}CH_2$

[Griffin, Bentrude and Johnson (1969).]

(d) $Ph_2P{-}PPh_2 + ArCH_2COOH \xrightarrow{180\,°C} ArMe + Ph_2P(=O)H + CO$

[Davidson, Sheldon and Trippett (1968).]

(e) $Me_2N\cdot + ROPPh_2 \longrightarrow Me_2NP(=O)Ph_2 + RP(=O)Ph_2$

[Davidson (1968).
See also Chapter 6, pp. 216–23.]

Problems Involving Synthesis

1 The following reaction scheme is similar to that used in one route to optically active phosphines and phosphonium salts (*J. Amer. Chem. Soc.*, 1968, **90**, 4842). Suggest the reagents and conditions required for carrying out each step.

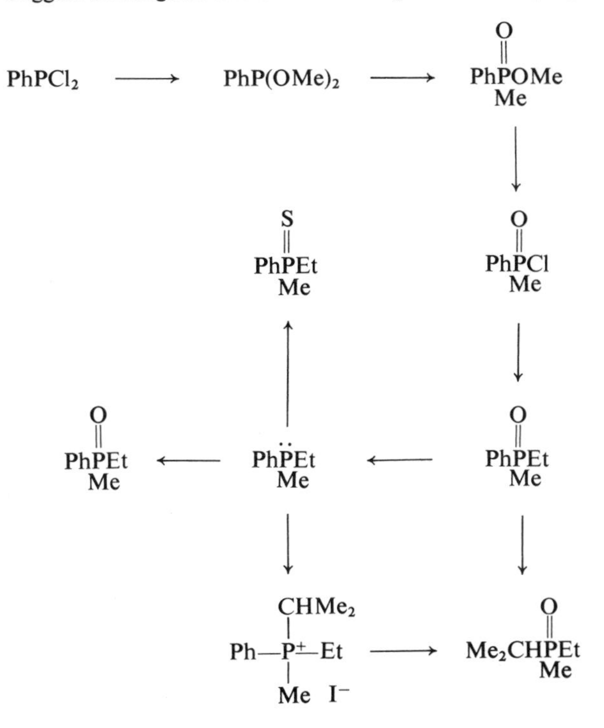

2 How would you prepare:

(a) $Ph_2\overset{O}{\overset{\|}{P}}Me$ from Ph_2POMe

(b) $Ph_2P(=O)Et$ from Ph_2POMe

(c) Et_3P from PCl_3

(d) $(Me_2N)_3P$ from PCl_3

(e) $Ph_2P(=O)Et$ from Ph_3P (via $Ph_3P^+EtBr^-$)

(f) $Et_2P(=S)—P(=S)Et_2$ from $Cl_3P{=}S$

(g) $Ph_2P(=O)CHPh.NHPh$ from $Ph_2P(=O)H$

(h) $Ph_3P^+C^-HCOOEt$ from Ph_3P

(i) 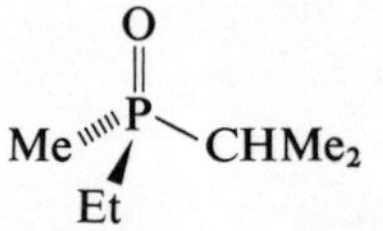 Me(O=)P(Et)CHMe₂ (chiral, as drawn) from $Me_2CH—P^+(Ph)(Et)Me\ I^-$ (chiral, as drawn)

(j) $(EtO)_2P(=O)CH_2Ph$ from $(EtO)_3P$

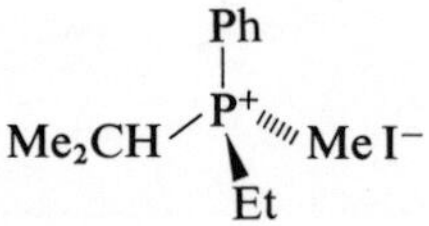

(k) 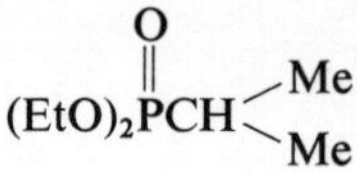 $(EtO)_2P(=O)CH(Me)Me$ from $(EtO)_3P$

(l) $Ph_3P^+C^-HCOPh$ (i) from Ph_3P and (ii) from $Ph_3P^+—C^-H_2$

(m) $PhCH{=}CHPh$ from $Ph_3P^+—CH_2PhBr^-$
(*cis* and *trans*)

(n) $Ph_3P^+—C(Me)(Me)—Me\ I^-$ from Ph_3P

(o) 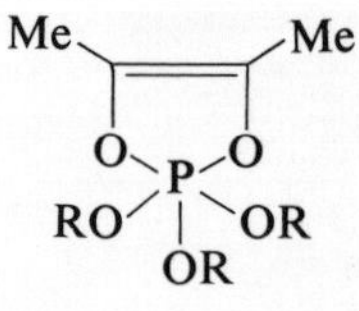 cyclic phosphorane: MeC=CMe ring closed through O–P–O, P bearing three OR groups, from $MeCO.COMe$

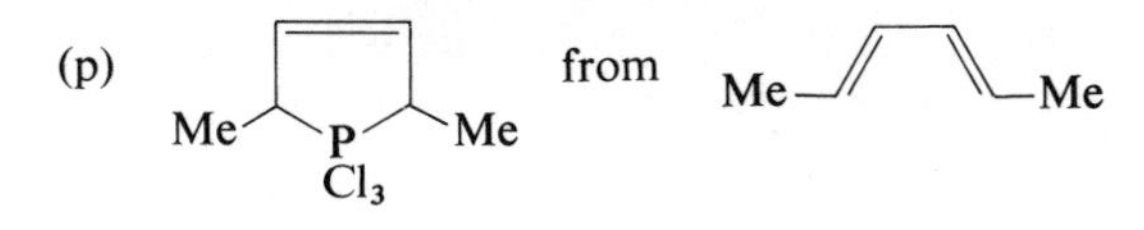

(q) from PCl_3

3 Predict the products of the following reactions.

(a) $Ph_2PCl + MeOH \xrightarrow{Et_3N} \xrightarrow{EtI}$; ↓ heat

(b) $Et_2PCl_3 + n MeOH \longrightarrow$

(i) $n = 1$; (ii) $n = 2$.

(c) $(EtO)_3P + MeCO.COMe \longrightarrow$

(d) $R\overset{O}{\overset{\|}{P}}(OH)_2 + 2PCl_5 \longrightarrow$

(e) $Ph_3PCl_2 + 2PhNH_2 \longrightarrow \xrightarrow{HBr}$; ↓ $H.C{\equiv}C.COOMe$

(f) $Ph_2PCl_3 + 4NH_3 \longrightarrow$

4 Outline a method of synthesis for each of the following ylids, paying particular attention to the reagents and conditions used in each case.

(a) $Ph_3P^+{-}C^-H{-}Ph$

(b) $Ph_3P^+{-}C^-H{-}COOEt$

(c) $Ph_3P^+{-}\bar{C}(Me)(Me)$

(d) $Ph_3P^+{-}\bar{C}(COOEt)(COOEt)$

(e) $Ph_3P^+{-}C^-(Me){-}COPh$

(f) $Me_3P^+{-}C^-H_2$

5 Describe how you would synthesize the following olefins, bearing in mind the availability of starting materials and the reactivity of intermediates.

(a) $Ph.CH{=}C(Me)_2$

(b) $Ph.CH{=}CH.CH{=}CH.Ph$

(c) *cis* $RCH{=}CHPh$

(d) methylenecyclohexane (cyclohexane ring with $=CH_2$)

(e) $CH_2{=}C(Ph)_2$

Appendix

Table A1 Characteristic infrared absorption frequencies of phosphorus–X bonds

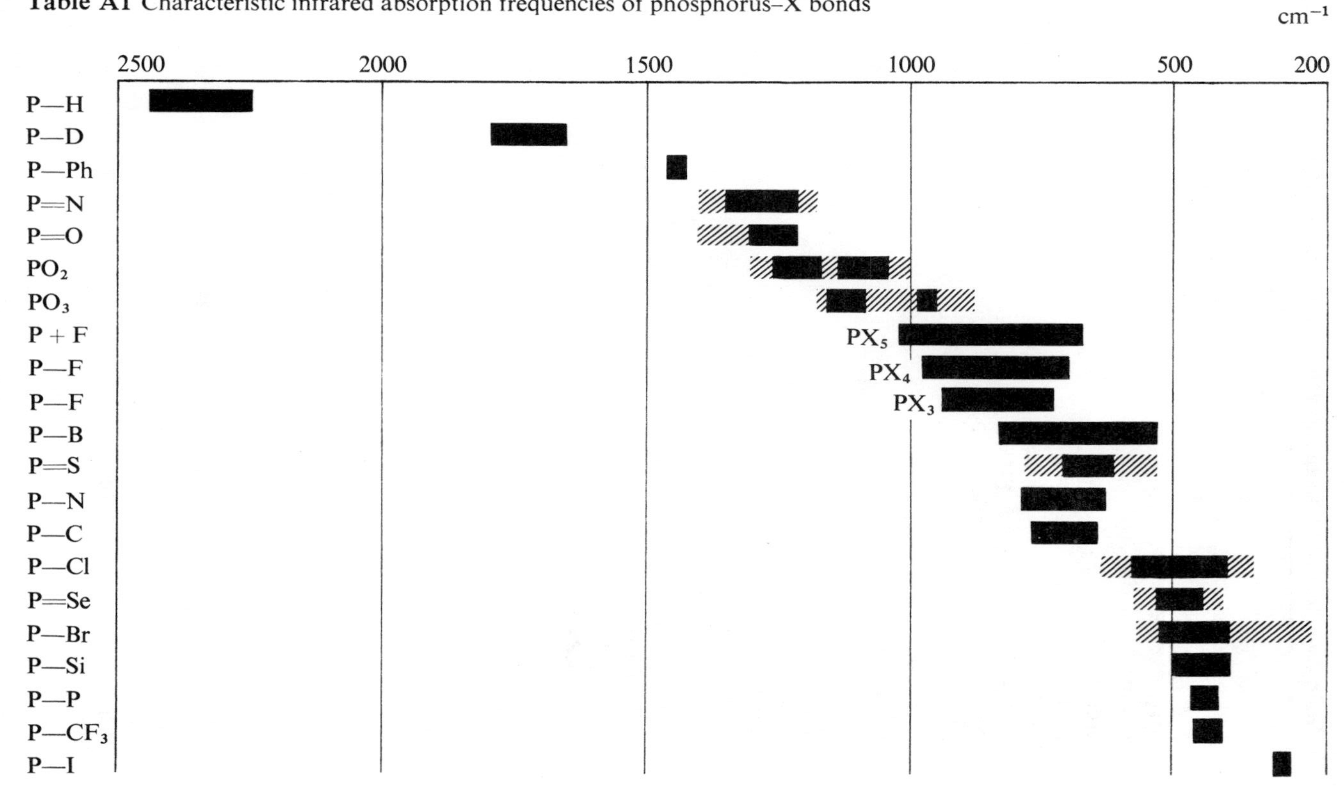

Table A2 Chemical shifts of some organophosphorus compounds

Compound	*Chemical shift in p.p.m. from* $(HO)_3PO$
R_3P	
PH_3	+240
Me_3P	+61
nBu_3P	+32
Ph_3P	+7
$MePCl_2$	−191
$MeP(OR)_2$	−175
P_4	+460
Cl_3P	−219
$(MeO)_3P$	−140
$(Me_2N)_3P$	−122
$R_3P{=}O$	
$Et_3P{=}O$	−48
$Bu_3P{=}O$	−43
$Me_2P(O)OEt$	−50
$Cl_3P{=}O$	−3
$(EtO)_2P(O)Cl$	−2·8
$(HO)_3P{=}O$	0
$(O^-)_3P{=}O$	−6
$(Me_2N)_3P{=}O$	−24
$R_3P{=}S$	
$Et_3P{=}S$	−54
$Br_3P{=}S$	+112
$Cl_3P{=}S$	−30
$(MeO)_3P{=}S$	−73
$(Et_2N)_3P{=}S$	−77
PX_5	
$(MeO)_3P(O\text{–}CR{=}CR\text{–}O)$	+50
$PhPF_4$	+52
PCl_5	+80
PBr_5	+101

Table A2 – *continued*

Compound	*Chemical shift in p.p.m. from* $(HO)_3PO$
$R_4P^+X^-$	
$Me_4P^+Br^-$	–25
$Ph_3P^+MeBr^-$	–23
$Ph_4P^+I^-$	–22
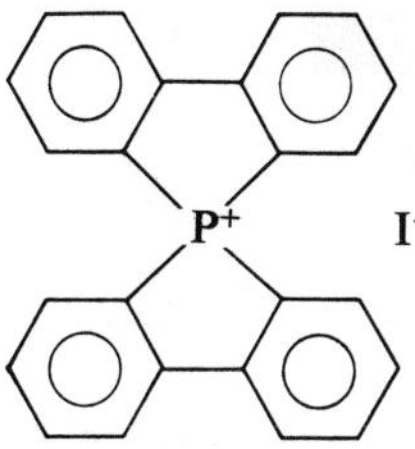 (bis-biphenylylene phosphonium structure) P^+ I^-	–24

Table A3 Direct phosphorus–hydrogen coupling constants

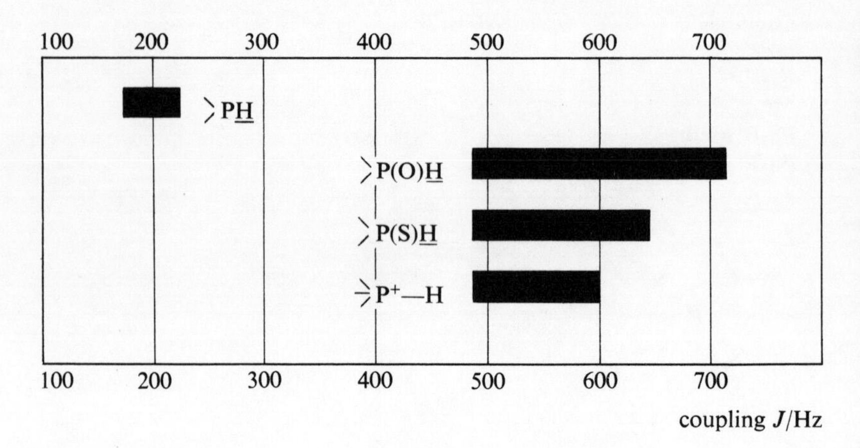

Table A4 Proton chemical shifts in phosphorus compounds

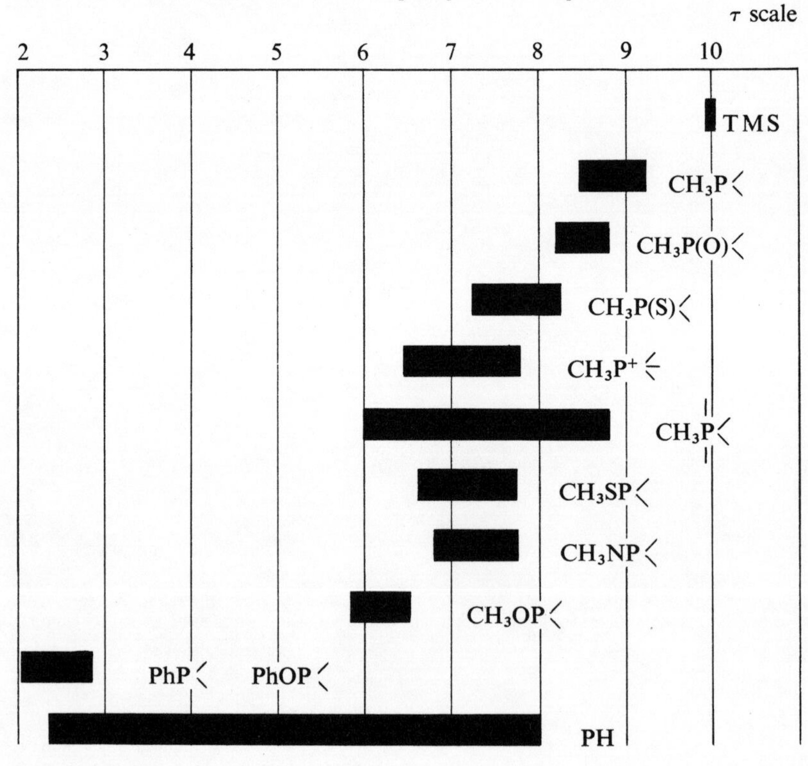

Table A5 Phosphorus–hydrogen coupling constants

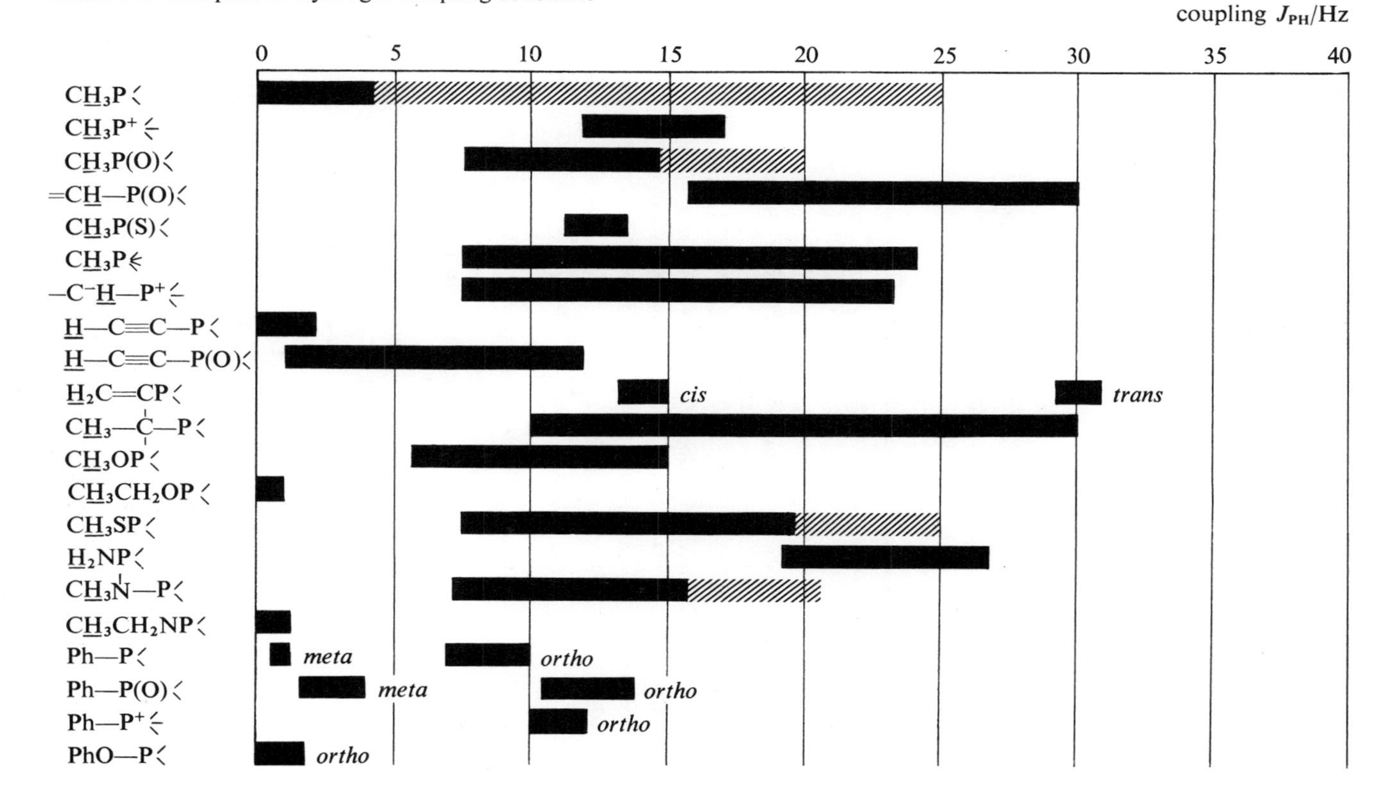

References

ABRAMOVITCH, R. A., and DAVIS, B. A. (1964), *Chemical Reviews*, vol. 64, pp. 149–85.

ANDERSON, A. G., and FREENOR, F. J. (1964), *Journal of the American Chemical Society*, vol. 86, pp. 5037–8.

BARTHEL, W. F., ALEXANDER, B. H., GIANG, P. A., and HALL, S. A. (1955), *Journal of the American Chemical Society*, vol. 77, pp. 2424–7.

BENTRUDE, W. G., HARGIS, J. H., and RUSEK, P. E. (1969), *Chemical Communications*, pp. 296–7.

BENTRUDE, W. G., and WIELESEK, R. A. (1969), *Journal of the American Chemical Society*, vol. 91, pp. 2406–7.

BERGELSON, L. D., VAVER, V. A., BARSUKOV, L. I., and SCHEMYAKIN, M. M. (1963), *Izvestiya Akademii Nauk SSSR, Otdelenie Khimicheskihk Nauk*, pp. 1134–6; *Chemical Abstracts*, vol. 59, p. 8607d.

BESTMANN, H. J., and KUNSTMANN, R. (1969), *Chemische Berichte*, vol. 102, pp. 1816–32.

BESTMANN, H. J., and SCHULZ, H. (1964), *Annalen der Chemie*, vol. 674, pp. 11–17.

BIRUM, G. H., and MATTHEWS, C. M. (1967), *Chemical Communications*, pp. 137–8.

BLOOM, S. M., BUCKLER, S. A., LAMBERT, R. F., and MERRY, E. V. (1970), *Chemical Communications*, pp. 870–71.

BOCHWIC, B., and FRANKOWSKI, A. (1968), *Tetrahedron*, vol. 24, pp. 6653–62.

BROWN, G. W., COOKSON, R. C., STEVENS, I. D. R., MAK, T. C. W., and TROTTER, J. (1964), *Proceedings of the Chemical Society*, pp. 87–8.

CORFIELD, J. R., HARGER, M. J. P., SHUTT, J. R., and TRIPPETT, S. (1970), *Journal of the Chemical Society*, series C, pp. 1855–60.

COSTAIN, C. C., and SUTHERLAND, G. B. B. M. (1952), *Journal of Physical Chemistry*, vol. 56, pp. 321–4.

DAVIDSON, R. S. (1968), *Tetrahedron Letters*, pp. 3029–31.

DAVIDSON, R. S., SHELDON, R. A., and TRIPPETT, S. (1968), *Journal of the Chemical Society*, series C, pp. 1700–703.

EGAN, W., TANG, R., ZON, G., and MISLOW, K. (1970), *Journal of the American Chemical Society*, vol. 92, pp. 1442–4.

FILLEUX-BLANCHARD, M. L., and MARTIN, M. G. J. (1970), *Comptes rendus hebdomadaires des séances de l'Academie des Sciences*, series C, vol. 270, pp. 1747–9.
GRIFFIN, C. E., BENTRUDE, W. G., and JOHNSON, G. M. (1969), *Tetrahedron Letters*, pp. 969–72.
GRAYSON, M., and GRIFFITH, E. J. (eds.) (1964), *Topics in Phosphorus Chemistry*, vol. 1, Interscience.
GRAYSON, M., and GRIFFITH, E. J. (eds.) (1965), *Topics in Phosphorus Chemistry*, vol. 2, Interscience.
HALL, L. A. R., STEPHENS, C. W., and DRYSDALE, J. J. (1957), *Journal of the American Chemical Society*, vol. 79, pp. 1768–9.
HAMER, N. K. (1968), *Chemical Communications*, p. 1399.
HARTLEY, S. B., HOLMES, W. S., JACQUES, J. K., MOLE, M. F., and MCCOUBREY, J. C. (1963), *Quarterly Reviews*, vol. 17, pp. 204–23.
HAWES, W., and TRIPPETT, S. (1968), *Chemical Communications*, pp. 577–8.
HOFFMANN, H., and DIEHR, H. J. (1962), *Tetrahedron Letters*, pp. 583–6.
HORNER, L., and WINKLER, H. (1964), *Tetrahedron Letters*, pp. 455–60.
HOUSE, H. O., and BABAD, H. (1963), *Journal of Organic Chemistry*, vol. 28, pp. 90–92.
HUDSON, R. F. (1965), *Structure and Mechanism in Organophosphorus Chemistry*, Academic Press.
HUISGEN, R., and WULFF, J. (1969), *Chemische Berichte*, vol. 102, pp. 1833–40.
KENNARD, O., MANN, F. G., WATSON, D. G., FAWCETT, J. K., and KERR, K. A. (1968), *Chemical Communications*, pp. 269–71.
KIRBY, A. J., and WARREN, S. G. (1967), *The Organic Chemistry of Phosphorus*, Elsevier.
LAMBERT, J. F., JACKSON, G. F., III and MUELLER, D. C. (1968), *Journal of the American Chemical Society*, vol. 90, pp. 6401–5.
MAGERLEIN, H., and MEYER, G. (1970), *Chemische Berichte*, vol. 103, pp. 2995–7.
MOORE, C. G., and TREGO, B. R. (1962), *Tetrahedron*, vol. 18, pp. 205–18.
MUKAIYAMA, T., NAMBU, H., and OKAMOTO (1962), *Journal of Organic Chemistry*, vol. 27, pp. 3651–4.
NEFEDOV, O. M., and MANAKOV, M. N. (1966), *Angewandte Chemie*, International edn, vol. 5, pp. 1021–38.
OBRYCKI, R., and GRIFFIN, C. E. (1968), *Journal of Organic Chemistry*, vol. 33, pp. 632–6.
PAULING, L. (1960), *The Nature of the Chemical Bond*, 3rd edn, Cornell University Press.
PETRELLIS, P., and GRIFFIN, C. W. (1968), *Chemical Communications*, pp. 1099–100.
PETROV, K. A., and NEIMYESHEVA, A. A. (1959), *Zhurnal Obsnichei Khimicheskikh*, vol. 29, pp. 1822–6.

PETROV, K. A., and NEIMYESHEVA, A. A. (1960), *Chemical Abstracts*, vol. 54, p. 8600l.
POSHKUS, A. C., and HERWEH, J. E. (1962), *Journal of Organic Chemistry*, vol. 27, pp. 2700–702.
REGITZ, M., ANSCHÜTZ, W., BARTZ, W., and LIEDHEGENER, A. (1968), *Tetrahedron Letters*, pp. 3171–4.
SCHMUTZLER, R. (1965), *Angewandte Chemie*, International edn, vol. 4, pp. 496–508.
SCHOMAKER, V., and STEVENSON, D. P. (1941), *Journal of the American Chemical Society*, vol. 63, pp. 37–40.
SCHÖNBERG, A., BROSOWSKI, K. H., and SINGER, E. (1962), *Chemische Berichte*, vol. 95, pp. 2144–54.
SCHWEIZER, E. E. (1964), *Journal of the American Chemical Society*, vol. 86, p. 2744.
SCHWEIZER, E. E., and LIGHT, K. K. (1964), *Journal of the American Chemical Society*, vol. 86, p. 2963.
SEYFERTH, D., and BURLITCH, J. M. (1963), *Journal of Organic Chemistry*, vol. 28, pp. 2463–4.
SHAW, M. A., and TEBBY, J. C. (1970), *Journal of the Chemical Society*, series C, pp. 5–9.
SNYDER, J. P., and BESTMANN, H. J. (1970), *Tetrahedron Letters*, pp. 3317–20.
SPEZIALE, A. J., and TUNG, C. C. (1963), *Journal of Organic Chemistry*, vol. 28, pp. 1353–7.
STAUDINGER, H., and MEYER, J. (1919), *Helvetica Chimica Acta*, vol. 2, pp. 635–46.
STOCKEL, R. F. (1968), *Chemical Communications*, pp. 1594–5.
TRIPPETT, S. (ed.) (1971), *Specialist Periodical Reports, Organophosphorus Chemistry*, vol. 2.
VAN WAZER, J. R. (1958), *Phosphorus and Its Compounds*, vol. 1, Interscience.
WESTHEIMER, H. F. (1968), *Accounts of Chemical Research*, vol. 1, pp. 71–8.
WESTON, R. E. (1954), *Journal of the American Chemical Society*, vol. 76, pp. 2645–8.
WILSON, I. F., and TEBBY, J. C. (1970), *Tetrahedron Letters*, pp. 3369–70.
WITTIG, G., EGGERS, H., and DUFFNER, P. (1958), *Annalen der Chemie*, vol. 619, pp. 10–27.

Bibliography

General

Books

A. J. KIRBY and S. G. WARREN, *The Organic Chemistry of Phosphorus*, Elsevier, 1967.
An advanced text covering the whole range of organophosphorus chemistry from a mechanistic viewpoint.

G. M. KOSSOLAPOFF, *Organophosphorus Compounds*, Wiley, 1950.
The first book on organophosphorus chemistry; still provides useful information.

K. SASSE, *Phosphorus Compounds*, vol. 12 of E. Müller (ed.), *Methoden der Organischen Chemie* (Houben-Weyl), Thieme, 1963, parts 1 and 2.
This comprehensive text gives a complete coverage of the methods of synthesis and reactions of organophosphorus compounds.

J. R. VAN WAZER, *Phosphorus and Its Compounds*, Interscience, 1958, 2 vols.
Vol. 1 deals with both the organic and inorganic chemistry of phosphorus compounds, while vol. 2 covers the commercial aspects of phosphorus.

Reviews

P. C. CROFTS, 'Compounds containing C—P bonds', *Quarterly Reviews*, vol. 12, 1958, pp. 341–66.

M. GRAYSON and E. J. GRIFFITH (eds.), *Topics in Phosphorus Chemistry*, Interscience.
This annual publication has so far reached vol. 7 (1970). It presents review articles on selected aspects of phosphorus chemistry.

L. HORNER, 'Preparative phosphorus chemistry', *Fortschritte der Chemischen Forschung*, vol. 7, 1966, pp. 1–61.

S. TRIPPETT (ed.), *Specialist Periodical Reports, Organophosphorus Chemistry*, The Chemical Society.
This consists of an annual survey of all the literature concerning organophosphorus compounds. Vol. 1 appeared in 1970.

Pure and Applied Chemistry, vol. 9, 1964, pp. 193ff.
This issue reports in full the main lectures from the IUPAC conference on organophosphorus held in Heidelberg in 1964.

Introduction
Books

M. BOAS, *Robert Boyle and Eighteenth Century Chemistry*, Cambridge University Press, 1958, pp. 226–7.

D. I. DUVEEN and H. S. KLICKSTEIN, *A Bibliography of the Works of Lavoisier*, Dawson & Weil, 1954, pp. 21–2, 42.

H. GUERLAC, *Lavoisier – The Crucial Year*, Cornell University Press, 1961.

J. NEWTON FRIEND, *Man and the Chemical Elements*, Griffiths, 1951, pp. 76–80.

J. READ, *Humour and Humanism in Chemistry*, Bell, 1947, pp. 168–73.

J. R. VAN WAZER, *Phosphorus and Its Compounds*, vol. 2, Interscience, 1958. Vol. 2 covers the commercial uses of phosphorus, including the manufacture of detergents, fertilizers, matches, etc.

Review

F. KRAFFT, 'Phosphorus – from elemental light to chemical element', *Angewandte Chemie*, International edn, vol. 8, 1969, pp. 660–71.
Describes the discovery of phosphorus in the seventeenth century.

1 Structure and bonding
Books

E. CARTMELL and G. W. A. FOWLES, *Valency and Molecular Structure*, 3rd edn, Butterworths, 1966.
Provides a good, simple introduction to quantum theory.

C. A. COULSON, *Valence*, 2nd edn, Clarendon Press, 1961.
Covers quantum theory in more detail and provides a full coverage of the bonding concepts in Chapter 1.

J. W. EMSLEY, J. FEENEY and L. H. SUTCLIFFE, *High Resolution Nuclear Magnetic Resonance Spectroscopy*, vol. 2, Pergamon, 1966.
An advanced text on nuclear magnetic resonance spectroscopy, which briefly covers phosphorus-31 spectra on pp. 1052–77.

R. F. HUDSON, *Structure and Mechanism in Organophosphorus Chemistry*, Academic Press, 1965.
Provides an excellent coverage of bonding and mechanism in the whole range of phosphorus chemistry; particularly useful in the way it relates these concepts to the chemistry of the compounds concerned.

A. J. KIRBY and S. G. WARREN, *The Organic Chemistry of Phosphorus*, Elsevier, 1967.
Pages 1–11 include a brief discussion of bonding and structure in phosphorus compounds.

J. R. VAN WAZER, *Phosphorus and Its Compounds*, vol. 1, Interscience, 1958.
Pages 1–60 include a detailed discussion of phosphorus atomic structure followed by bonding and some spectroscopy.

Reviews

D. E. C. CORBRIDGE, 'Structural chemistry of phosphorus compounds', in M. Grayson and E. J. Griffith (eds.), *Topics in Phosphorus Chemistry*, vol. 3, Interscience, 1966, pp. 57 ff.
Contains a large quantity of information on structural parameters (bond lengths, angles, etc.) of both inorganic and organic phosphorus compounds.

S. B. HARTLEY, W. S. HOLMES, J. K. JACQUES, M. F. MOLE and J. C. MCCOUBREY, 'Thermochemical properties of phosphorus compounds', *Quarterly Reviews*, vol. 17, 1963, pp. 204–23.

R. F. HUDSON, 'The nature of chemical bonding in organophosphorus compounds', *Pure and Applied Chemistry*, vol. 9, 1964, pp. 371–86.
Covers similar ground to *Structure and Mechanism in Organophosphorus Chemistry*, but in less detail.

N. L. PADDOCK, 'Structure and reactions in phosphorus chemistry', *Royal Institute of Chemistry Lecture Series*, 1962, no. 2.

R. SCHMUTZLER, 'Chemistry and stereochemistry of fluorophosphoranes', *Angewandte Chemie*, International edn, vol. 4, 1965, pp. 496–508.
The preparation and chemistry of phosphorus(V) fluorine compounds, together with their structures and n.m.r. spectra, are discussed.

d-*Orbitals*

J. E. BISSEY, 'Some aspects of d-orbital participation in phosphorus and silicon chemistry', *Journal of Chemical Education*, vol. 44, 1967, pp. 95–100.
Some of the evidence for the involvement of d-orbitals in the bonding of phosphorus and silicon compounds is briefly discussed, with particular emphasis on bonding in the phosphazenes.

K. A. R. MITCHELL, 'The use of outer d-orbitals in bonding', *Chemical Reviews*, vol. 69, 1969, pp. 157–78.
An excellent critical review of the evidence available for, and the theory behind, the involvement of d-orbitals in the bonding of the second-row elements. Examples are taken mainly from the compounds of phosphorus and sulphur.

N. L. PADDOCK, 'Phosphonitrilic derivatives and related compounds', *Quarterly Reviews*, vol. 18, 1964, pp. 168–210.
See especially pp. 168, 180–210.

L. M. VENANZI, 'd_σ and d_π Bonding in phosphine complexes', *Chemistry in Britain*, vol. 4, 1968, pp. 162–7.

Methods of construction of models of d-orbitals can be found in B. E. Douglas, 'A simple model of d-orbitals', *Journal of Chemical Education*, vol. 41, 1964, p. 40; D. G. Nicholson, 'Simplified d-orbital models assist in teaching coordination concepts', *Journal of Chemical Education*, vol. 42, 1965, pp. 148–9; A. F. Kapanan, 'Cardboard orbital domain models', *Journal of Chemical Education*, vol. 43, 1966, pp. 412–13; E. J. Barrett, 'Models illustrating d-orbitals involved in multiple bonding', *Journal of Chemical Education*, vol. 44, 1967, pp. 146–7.

Spectroscopy

D. E. C. CORBRIDGE, 'The infrared spectra of phosphorus compounds', in M. Grayson and E. J. Griffith (eds.), *Topics in Phosphorus Chemistry*, vol. 6, Interscience, 1969, pp. 235–365.
Contains much up-to-date information on infrared absorption of phosphorus compounds.

R. A. Y. JONES and A. R. KATRITZKY, 'Nuclear magnetic resonance of phosphorus', *Angewandte Chemie*, 1962, vol. 74, pp. 60–68.
An early review of phosphorus-31 nuclear magnetic resonance.

G. MAVEL, 'Studies of phosphorus compounds using the magnetic resonance spectra of nuclei other than phosphorus-31', in J. W. Emsley, J. Feeney and L. H. Sutcliffe (eds.), *Progress in Nuclear Magnetic Resonance Spectroscopy*, vol. 1, Pergamon, 1966, pp. 260–373.
A good advanced review of phosphorus-31 and hydrogen-1 nuclear magnetic resonance spectroscopy of phosphorus compounds.

J. C. TEBBY, 'Physical methods', *Specialist Periodical Reports, Organophosphorus Chemistry*, 1970, vol. 1, pp. 273–322.
Reviews the application of physical methods to organophosphorus chemistry for 1968–9.

J. R. VAN WAZER, J. H. LETCHER, V. MARK, C. H. DUNGAN and M. M. CRUTCHFIELD, '^{31}P Nuclear magnetic resonance', in M. Grayson and E. J. Griffith (eds.), *Topics in Phosphorus Chemistry*, vol. 5, Interscience, 1968.
Covers the whole range of phosphorus-31 nuclear magnetic resonance and contains much useful information on chemical shifts, coupling constants, etc.

2 Trivalent-phosphorus compounds

Books

R. F. HUDSON, *Structure and Mechanism in Organophosphorus Chemistry*, Academic Press, 1965.
Following a good introduction to the theory of nucleophilicity, the reactions of trivalent phosphorus are discussed from a mechanistic viewpoint on pp. 90–198.

A. J. KIRBY and S. G. WARREN, *The Organic Chemistry of Phosphorus*, Elsevier, 1967.
Provides good coverage of the reactions of trivalent phosphorus from a mechanistic standpoint on pp. 32–132, 236–47.

L. D. QUIN, 'Trivalent phosphorus compounds as dienophiles', in J. Hamer (ed.), *1,4-Cycloaddition Reactions*, Academic Press, 1967, pp. 47–96.
Gives an advanced coverage of the 1,4-cycloaddition reactions of dienophilic phosphorus compounds.

K. SASSE, *Phosphorus Compounds*, vol. 12 of E. Müller (ed.), *Methoden der Organischen Chemie* (Houben-Weyl), Thieme, 1963, part 1, pp. 17–79.

Reviews

B. A. ARBUSOV, 'The Michaelis–Arbusov and Perkow reactions', *Pure and Applied Chemistry*, vol. 9, 1964, pp. 307–35.

K. D. BERLIN, T. H. AUSTIN, M. PETERSON and M. NAGABHUSHANAM, 'Nucleophilic displacement reactions on phosphorus halides and esters by Grignard and lithium reagents', in M. Grayson and E. J. Griffith (eds.), *Topics in Phosphorus Chemistry*, vol. 1, Interscience, 1964, pp. 17–55.
Electrophilic reactions of both phosphorus(III) and phosphorus(V) halides are reviewed.

J. I. G. CADOGAN, 'Oxidation of tervalent organic compounds of phosphorus', *Quarterly Reviews*, vol. 16, 1962, pp. 208–39.

E. FLUCK, 'Phosphorus nitrogen chemistry', in M. Grayson and E. J. Griffith (eds.), *Topics in Phosphorus Chemistry*, vol. 4, Interscience, 1967, pp. 291–481.
An advanced review covering the preparation and reactions of the whole range of phosphorus–nitrogen compounds.

R. G. HARVEY and E. R. DESOMBRE, 'Michaelis–Arbusov and related reactions', in M. Grayson and E. J. Griffith (eds.), *Topics in Phosphorus Chemistry*, vol. 1, Interscience, 1964, pp. 57–112.
A review of oxidation reactions of trivalent-phosphorus esters in general and the Michaelis–Arbusov reaction in particular.

H. HOFFMANN and H. J. DIEHR, 'Phosphonium salt formation of the second kind', *Angewandte Chemie*, International edn, vol. 3, 1964, pp. 737–46.
Reactions which involve the attack of trivalent phosphorus at halogen, rather than carbon, atoms are discussed.

L. HORNER, 'Preparation and properties of optically active tertiary phosphines', *Pure and Applied Chemistry*, vol. 9, 1964, pp. 225–44.
A useful review of this topic, although it contains a number of errors, particularly in diagrams, that are often confusing.

D. W. HUTCHINSON, 'Tervalent phosphorus acids', *Specialist Periodical Reports, Organophosphorus Chemistry*, vol. 1, 1970, pp. 80–97.

K. ISSLEIB, 'Synthesis of organophosphorus compounds by the reactions of *P*-substituted metal phosphines', *Pure and Applied Chemistry*, vol. 9, 1964, pp. 205–23.

L. MAIER, 'Preparation and properties of primary, secondary and tertiary phosphines', in F. A. Cotton (ed.), *Progress in Inorganic Chemistry*, vol. 5, Interscience, 1963, pp. 27–210.
A comprehensive review containing over five hundred references.

B. MILLER, 'Reactions between trivalent phosphorus derivatives and positive halogen sources', M. Grayson and E. J. Griffith (eds.), *Topics in Phosphorus Chemistry*, vol. 2, Interscience, 1965, pp. 133–200.
Reactions with activated halogen compounds (e.g. α-haloketones, *N*-chloroamines, sulphur halides), including the Perkov reaction, are covered.

F. RAMIREZ, 'Condensations of carbonyl compounds with phosphite esters', *Pure and Applied Chemistry*, vol. 9, 1964, pp. 337–69.

S. TRIPPETT, 'Phosphines', *Specialist Periodical Reports, Organophosphorus Chemistry*, vol. 1, 1970, pp. 1–21.

S. TRIPPETT, 'Halogenophosphines and related compounds', *Specialist Periodical Reports, Organophosphorus Chemistry*, vol. 1, 1970, pp. 55–60.
These two papers, and that of Hutchinson, review the chemistry of trivalent phosphorus for 1968–9.

3 The pentavalent state – compounds derived from PX_5

Books

H. R. ALLCOCK, *Heteroatom Ring Systems and Polymers*, Academic Press, 1967.
Includes on pp. 128–40 a brief discussion of phosphonitrilic compounds, with emphasis on the properties of their polymers.

A. J. KIRBY and S. G. WARREN, *The Organic Chemistry of Phosphorus*, Elsevier, 1967.
Covers the reactions of phosphorus(V) compounds from a mechanistic viewpoint on pp. 262–70, 274–353.

G. M. KOSSOLAPOFF, *Organophosphorus Compounds*, Wiley, 1950, pp. 98–277.

K. SASSE, *Phosphorus Compounds*, vol. 12 of E. Müller (ed.), *Methoden der Organischen Chemie* (Houben-Weyl), Thieme, 1963, part 1, pp. 125–81, 193–620; part 2, pp. 1–994.

Reviews

M. Becke-Goehring, 'The chemistry of phosphorus pentachlorides', *Fortschritte der Chemischen Forschung*, vol. 10, 1968, pp. 207–37.
Discusses the structure of phosphorus pentachloride and its reactions with nitrogen compounds, including those giving phosphonitrilics.

M. Becke-Goehring and E. Fluck, 'The route from PCl_5 to phosphonitrilic chlorides', *Angewandte Chemie*, vol. 74, 1962, pp. 382–6.
The reactions of phosphorus pentachloride with amino groups that give phosphonitrilic chlorides are discussed. (Thirty-two references.)

F. Cramer, 'Preparation of esters, amides and anhydrides of phosphoric acid', *Angewandte Chemie*, vol. 72, 1960, pp. 236–49.
A review with 160 references covering the range of phosphoric acid derivatives.

G. O. Doak and L. D. Freedman, 'The structure and properties of dialkylphosphonates', *Chemical Reviews*, vol. 61, 1961, pp. 31–44.

E. Fluck, 'Phosphorus nitrogen chemistry', in M. Grayson and E. J. Griffith (eds.), *Topics in Phosphorus Chemistry*, vol. 4, Interscience, 1967, pp. 327–407.

D. W. Hutchinson, 'Quinquevalent phosphorus acids', *Specialist Periodical Reports, Organophosphorus Chemistry*, vol. 1, 1970, pp. 98–141.

R. Keat and R. A. Shaw, 'Phosphazenes', *Specialist Periodical Reports, Organophosphorus Chemistry*, vol. 1, 1970, pp. 214–39.

F. A. Lichtenhaler, 'The chemistry and properties of enol-phosphates', *Chemical Reviews*, vol. 61, 1961, pp. 607–49.
Includes a coverage of the Perkov reaction.

S. Ohashi, 'Condensed phosphates containing other oxo acid anions', in M. Grayson and E. J. Griffith (eds.), *Topics in Phosphorus Chemistry*, vol. 1, Interscience, 1964, pp. 189–240.

N. L. Paddock, 'Phosphonitrilic derivatives and related compounds', *Quarterly Reviews*, vol. 18, 1964, pp. 168–210.

D. S. Payne, 'The chemistry of phosphorus halides', in M. Grayson and E. J. Griffith (eds.), *Topics in Phosphorus Chemistry*, vol. 4, Interscience, 1967, pp. 117–44.
A review of the phosphorus pentahalides which is almost entirely inorganic.

F. Ramirez, 'Condensation of carbonyl compounds with phosphite esters', *Pure and Applied Chemistry*, vol. 9, 1964, pp. 337–69.

C. D. Schmulbach, 'Phosphonitrile polymers', in F. A. Cotton (ed.), *Progress in Inorganic Chemistry*, vol. 4, Interscience, 1962, pp. 275–380.
A discussion of the structures, preparation and properties of phosphonitrilic compounds. (194 references.)

R. Schmutzler, 'Chemistry and stereochemistry of the fluorophosphoranes', *Angewandte Chemie*, International edn, vol. 4, 1965, pp. 496–508.

S. TRIPPETT, 'Quinquevalent phosphorus compounds', *Specialist Periodical Reports, Organophosphorus Chemistry*, vol. 1, 1970, pp. 40–55.

S. TRIPPETT, 'Halogeno-phosphoranes', *Specialist Periodical Reports, Organophosphorus Chemistry*, vol. 1, 1970, pp. 60–64.
These two papers by Trippett, together with those of Hutchinson, and Keat and Shaw, review the chemistry of phosphorus(V) compounds for 1968–9.

M. WEBSTER, 'Addition compounds of the group V pentahalides', *Chemical Reviews*, vol. 66, 1966, pp. 87–118.
Adducts of the pentahalides of phosphorus, arsenic and antimony with a large number of donor molecules are discussed.

F. H. WESTHEIMER, 'Pseudorotation in the hydrolysis of phosphate esters', *Accounts of Chemical Research*, vol. 1, 1968, pp. 71–8.
An excellent review of cyclic-ester hydrolysis by a leading authority in the field.

4 The pentavalent state – compounds derived from phosphonium salts

Books

A. W. JOHNSON, *Ylid Chemistry*, Academic Press, 1966.
Gives an excellent, in-depth coverage of the chemistry of ylids, and related compounds, of phosphorus, sulphur, nitrogen and arsenic.

A. J. KIRBY and S. G. WARREN, *The Organic Chemistry of Phosphorus*, Elsevier, 1967.
The Wittig reaction, other reactions of phosphorus ylids and reactions of phosphonium salts are dealt with on pp. 184–204, 221–31, 250–61.

G. M. KOSSOLAPOFF, *Organophosphorus Compounds*, Wiley, 1950, pp. 78–97.

L. L. MULLER and J. HAMER, *1,2-Cycloaddition Reactions*, Interscience, 1967.
The Wittig and related reactions are discussed on pp. 305–32. Some useful information on yields and products of reactions, and spectral data are included.

K. SASSE, *Phosphorus Compounds*, vol. 12 of E. Müller (ed.), *Methoden der Organischen Chemie* (Houben-Weyl), Thieme, 1963, part 1, pp. 79–123.

U. SCHÖLLKOPF, 'The Wittig reaction', in W. Foerst (ed.), *Newer Methods of Preparative Organic Chemistry*, vol. 3, Academic Press, 1964, pp. 111–50.

Reviews

H. J. BESTMANN, 'New reactions of phosphoranes and their preparative possibilities', *Pure and Applied Chemistry*, vol. 9, 1964, pp. 285–306.

H. J. BESTMANN, 'New reactions of the alkylidene phosphoranes and their preparative uses', *Angewandte Chemie*, International edn, vol. 4, 1965, pp. 583–7, 645–60, 830–38.
The whole range of phosphonium ylid reactions known in 1965 is discussed, with particular regard to their use in synthesis.

E. FLUCK, 'Phosphorus nitrogen chemistry', in M. Grayson and E. J. Griffith (eds.), *Topics in Phosphorus Chemistry*, vol. 4, Interscience, 1967, pp. 409–27.
Covers the chemistry of phosphine imides.

L. HORNER, 'Preparative phosphorus chemistry', *Fortschritte der Chemischen Forschung*, vol. 7, 1966, pp. 1–61.
A number of reactions that involve phosphorus compounds and have preparative use are discussed. Particular emphasis is placed on those reactions producing olefins.

S. TRIPPETT, 'Phosphonium salts', *Specialist Periodical Reports, Organophosphorus Chemistry*, vol. 1, 1970, pp. 21–33.

S. TRIPPETT, 'Ylids and related compounds', *Specialist Periodical Reports, Organophosphorus Chemistry*, vol. 1, 1970, pp. 176–208.
Reviews the reactions of phosphonium salts and phosphonium ylids appearing in the literature during 1968–9.

The Wittig reaction

H. FREYSCHLAG, H. GRASSNER, A. NÜRRENBACH, H. POMMER, W. REIF and W. SARNECKI, 'Formation and reactivity of phosphonium salts in the vitamin A series', *Angewandte Chemie*, International edn, vol. 4, 1965, pp. 287–91.
Deals with the formation of phosphonium salts containing conjugated systems.

S. TRIPPETT, 'The Wittig reaction', in R. A. Raphael, E. C. Taylor and H. Wynberg (eds.), *Advances in Organic Chemistry*, vol. 1, Interscience, 1960, pp. 83–102.
A review of the Wittig reaction covering its mechanism, applications and preparative procedure.

S. TRIPPETT, 'The Wittig reaction', *Quarterly Reviews*, vol. 17, 1963, pp. 406–40.
Includes a discussion of the preparation and reactions of phosphonium ylids as well as the mechanism of the Wittig reaction.

S. TRIPPETT, 'The Wittig reaction', *Pure and Applied Chemistry*, vol. 9, 1964, pp. 255–69.
The mechanism of the Wittig reaction.

G. WITTIG, 'The origin and development of the chemistry of phosphine-alkylenes', *Angewandte Chemie*, vol. 68, 1956, pp. 505–8.
An early review of the Wittig reaction.

G. WITTIG, 'Variations on a theme of Staudinger', *Pure and Applied Chemistry*, vol. 9, 1964, pp. 245–54.
A brief outline of the Wittig reaction up to 1964.

5 Phosphorus in biological processes

Books

J. D. BU'LOCK, *The Biosynthesis of Natural Products*, McGraw-Hill, 1965.
This paperback is a good introduction to the more specific processes of biosynthesis in plants and the methods of obtaining experimental information.

E. CHARGAFF and J. N. DAVIDSON (eds.), *The Nucleic Acids – Chemistry and Biology*, 3 vols., Academic Press, 1955, 1955, 1960.
The classic text on nucleic acids. Consists of a series of specialist chapters covering the whole range of nucleic acid chemistry and biology. Vols. 1 and 2 include work up to 1953. Vol. 3 supplements vols. 1 and 2 up to 1960. (See also J. N. Davidson and W. E. Cohn (eds.), *Progress in Nucleic Acid Research and Molecular Biology*.)

E. HARBERS, *Introduction to Nucleic Acids*, Reinhold, 1968.
A fairly advanced text covering the whole range of nucleic acid chemistry and biochemistry.

D. O. JORDAN, *The Chemistry of Nucleic Acids*, Butterworths, 1960.
A well-written account of the chemistry and structure of nucleic acids. Includes a particularly good introduction to the chemical make-up of nucleotides and nucleosides.

H. G. KHORANA, *Some Recent Developments in the Chemistry of Phosphate Esters of Biological Interest*, Wiley, 1961.
Reviews the synthesis and chemistry of a range of biologically important phosphate esters, by a recent Nobel prize winner.

A. L. LEHNINGER, *Bioenergetics*, Benjamin, 1965.
A really excellent introduction to the complexities of energy processes in living organisms. A whole range of subjects are dealt with within the framework of biological energy.

A. M. MICHELSON, *The Chemistry of Nucleosides and Nucleotides*, Academic Press, 1963.
Covers similar ground to D. O. Jordan, *The Chemistry of Nucleic Acids*, but in a more comprehensive manner.

V. R. POTTER, *Nucleic Acid Outlines*, vol. 1, *Structure and Metabolism*, vol. 2, *Function and Application*, Burgess, 1960, 1961.
Based on a series of lectures, this is an excellent account, from fairly simple beginnings, of nucleic acid function and chemistry up to the early 1960s. (For models of DNA and RNA, see V. R. Potter, 'DNA model kit', 1959.)

R. F. STEINER and R. F. BEERS, *Polynucleotides*, Elsevier, 1961.
Covers much the same ground as the books of Jordan and Michelson.

J. D. WATSON, *The Double Helix*, Weidenfeld & Nicholson, 1968 (Penguin, 1970).
'A personal account of the discovery of the structure of DNA' – this book deals with the human relationships as well as the science behind the unravelling of DNA structure.

Organophosphorus insecticides and nerve gases

D. F. HEATH, *Organo-Phosphorus Poisons*, Pergamon, 1961.
Covers similar ground to R. O'Brien, *Toxic Phosphorus Esters* and contains a very useful introduction which summarizes the subject as well.

R. O'BRIEN, *Toxic Phosphorus Esters*, Academic Press, 1960.
A fairly advanced text covering the chemistry, biology, biochemistry and physiological effects of phosphorus esters. Contains a large number of references, particularly on the biological aspects.

B. C. SAUNDERS, *Some Aspects of the Chemistry and Toxic Action of Organic Compounds Containing Phosphorus and Fluorine*, Cambridge University Press, 1957.
The first text in English to deal with the subject, it is now somewhat outdated, but still well worth reading as an introduction to the subject.

G. SCHRADER, *Die Entwicklung Neuer Insectiziden Phosphorsäure-Ester*, Chemie GMBH, 1963.
An excellent, typically Teutonic, systematic study of the organophosphorus insecticides known to 1963, by the father of this subject. For the factually minded everything is here: preparations, toxicities, physical and chemical properties, economics, etc.

T. F. WEST and J. E. HARDY, *Chemical Control of Insects*, 2nd edn, Chapman & Hall, 1961.
Contains a general discussion of the whole range of insecticides and their uses, and ch. 10 is devoted to organophosphorus compounds. The main insecticidal phosphorus components are dealt with in a similar way to G. Schrader, *Die Entwicklung Neuer Insectiziden Phosphorsäure-Ester* but in much less detail.

Reviews: nucleic acids and related systems

J. N. DAVIDSON and W. E. COHN (eds.), *Progress in Nucleic Acid Research and Molecular Biology*, 10 vols., Academic Press, 1963–70.
This annual publication contains articles, by authorities, on selected topics – in circumscribed areas – in the nucleic acid and related fields. It was originally intended as a continuing supplement to E. Chargaff and J. N. Davidson (eds.), *The Nucleic Acids – Chemistry and Biology*. The articles are advanced and aimed at a reader active in, or concerned with, the general field of nucleic acids.

A. KORNBERG, 'The biosynthesis of DNA', *Angewandte Chemie*, vol. 72, 1960, pp. 231–6.
Based on the Nobel lecture of Arthur Kornberg, who won the Prize in 1959 for his work on the biosynthesis of DNA.

A. R. TODD, 'Synthesis in the study of nucleotides', *Science*, vol. 127, 1958, pp. 787–92.
In an article based on his Nobel lecture of 1957, Lord Todd reviews his work on the phosphorylation of nucleosides.

T. L. V. ULBRICHT, 'Chemical synthesis of nucleotides', *Angewandte Chemie*, vol. 74, 1962, pp. 767–72.
Discusses the *in vitro* synthesis of nucleosides.

Reviews: organophosphorus insecticides and nerve gases

F. DUSPIVA, 'The mode of action of modern synthetic insecticides', *Angewandte Chemie*, vol. 66, 1954, pp. 541–51.
An early review, discussing the mechanism of action of both organophosphorus and halocarbon insecticides.

N. ENGELHARD, K. PRCHAL and M. NENNER, 'Acetylcholinesterase', *Angewandte Chemie*, International edn, vol. 6, 1967, pp. 615–26.
A review of acetylcholinesterase, which describes its mechanism of action and the effect of inhibiting substances.

G. SCHRADER, 'Chlorthion, a relatively nontoxic insecticide from the series of the thiophosphoric acid esters', *Angewandte Chemie*, vol. 66, 1954, pp. 265–7.

G. SCHRADER, 'Insecticidal phosphorus esters', *Angewandte Chemie*, vol. 69, 1957, pp. 86–90.

G. SCHRADER, 'New information on lesser toxic insecticides based on phosphorus esters', *Angewandte Chemie*, vol. 73, 1961, pp. 331–4.
These are all articles discussing the physical properties and toxicity, together with a little chemistry, of organophosphorus insecticides.

6 Radical and related reactions of phosphorus compounds

Books

A. J. KIRBY and S. G. WARREN, *The Organic Chemistry of Phosphorus*, Elsevier, 1967, pp. 158–81.
Provides a good coverage of free radical reactions involving phosphorus.

K. SASSE, *Phosphorus Compounds*, vol. 12 of E. Müller (ed.), *Methoden der Organischen Chemie* (Houben-Weyl), Thieme, 1963, part 1, pp. 182–93.

Reviews

J. I. G. CADOGAN, 'Phosphorus radicals', in G. H. Williams (ed.), *Advances in Free Radical Chemistry*, vol. 2, Logos, 1968, pp. 203–50.
Provides a good coverage of phosphorus radical reactions.

A. H. COWLEY, 'Chemistry of the phosphorus-phosphorus bond', *Chemical Reviews*, vol. 65, 1965, pp. 617–34.
Covers the chemistry of both diphosphines and cyclic polyphosphines.

A. H. COWLEY and R. P. PINNELL, 'Structures and reactions of cyclopolyphosphines', in M. Grayson and E. J. Griffith (eds.), *Topics in Phosphorus Chemistry*, vol. 4, Interscience, 1967, pp. 1–22.
Covers similar ground to that of L. Maier, 'Structure, preparation and reactions of cyclic polyphosphines'.

R. S. DAVIDSON, 'Radicals, photochemistry and deoxygenation reactions', *Specialist Periodical Reports, Organophosphorus Chemistry*, vol. 1, 1970, pp. 246–72.
Reviews photochemistry, deoxygenation and radical reactions involving phosphorus that appeared in the literature during 1968–9.

M. GRAYSON, 'Preparation of organophosphorus compounds from the element', *Pure and Applied Chemistry*, vol. 9, 1964, pp. 193–204.
This and M. M. Rauhut, 'Synthesis of organophosphorus compounds from elemental phosphorus' provide a good coverage of these reactions, which are of potential industrial importance.

M. HALMANN, 'Photochemical and radiation-induced reactions of phosphorus compounds', in M. Grayson and E. J. Griffith (eds.), *Topics in Phosphorus Chemistry*, vol. 4, Interscience, 1967, pp. 49–84.
Includes some discussion of phosphorus mass spectra.

J. E. HUHEEY, 'Chemistry of diphosphorus compounds', *Journal of Chemical Education*, vol. 40, 1963, pp. 153–8.
Provides a very brief review of compounds containing phosphorus–phosphorus bonds.

L. MAIER, 'Structure, preparation and reactions of cyclic polyphosphines', *Fortschritte der Chemischen Forschung*, vol. 8, 1967, pp. 1–60.
A comprehensive review of cyclic polyphosphines that thoroughly covers their structures, preparation and reactions.

O. M. NEFEDOV and M. N. MANAKOV, 'Inorganic, organometallic and organic analogues of carbenes', *Angewandte Chemie*, International edn, vol. 5, 1966, pp. 1021–38.
Includes a short discussion of 'phosphinidenes' (RP) on p. 1034.

M. M. RAUHUT, 'Synthesis of organophosphorus compounds from elemental phosphorus', in M. Grayson and E. J. Griffith (eds.), *Topics in Phosphorus Chemistry*, vol. 1, Interscience, 1964, pp. 1–16.

C. WALLING and M. S. PEARSON, 'Radical reactions of organophosphorus compounds', in M. Grayson and E. J. Griffith (eds.), *Topics in Phosphorus Chemistry*, vol. 3, Interscience, 1966, pp. 1–56.
A comprehensive review of phosphorus radical reactions and the reactions of organophosphorus compounds with other radicals.

E. WIBERG, M. VAN GHEMEN and G. MULLER-SCHIEDMAYER, 'New developments in the chemistry of polyphosphines', *Angewandte Chemie*, International edn, vol. 2, 1963, pp. 646–54.
Provides a brief review of both cyclic and acyclic polyphosphines.

Index